185-00

CRC Series in
CONTEMPORARY FOOD SCIENCE

Fergus M. Clydesdale, Series Editor
University of Massachusetts, Amherst

Published Titles:

America's Foods Health Messages and Claims:
Scientific, Regulatory, and Legal Issues
James E. Tillotson

New Food Product Development:
From Concept to Marketplace
Gordon W. Fuller

Forthcoming Titles:

Aseptic Processing and Packaging of Foods:
Food Industry Perspectives
Jarius David, V. R. Carlson, and Ralph Graves

Food Shelf Life Stability
N. A. Michael Eskin

Handbook of Foodborne Yeasts
Tibor Deak and Larry R. Beuchat

New Food Product Development

From Concept to Marketplace

Gordon W. Fuller

G.W. Fuller Associates, Ltd.
Montreal, Quebec

CRC Press
Boca Raton   Ann Arbor   London   Tokyo

Library of Congress Cataloging-in-Publication Data

Fuller, Gordon W.
    New food product development : from concept to marketplace /
Gordon W. Fuller.
      p. cm.
    Includes bibliographical references and index.
    ISBN 0-8493-8002-2
    1. Food--Marketing. 2. Food--Research. 3. New products-
-Marketing. 4. Product management. 5. Food industry and trade-
-Technological innovations. I. Title.
HD9000.5.F86   1994
641.3′0068′8—dc20   93-33273
                                                      CIP

© 1994 by CRC Press, Inc.

No claim to original U.S. Government works
International Standard Book Number 0-8493-8002-2
Library of Congress Card Number 93-33273
Printed in the United States of America   2 3 4 5 6 7 8 9 0
Printed on acid-free paper

DEDICATION

To my wife Joan, for both her diligent proofreading and her patience, and to my son Grahame, who acted as a user-friendly interface between me and the computer.

CRC Series in Contemporary Food Science

Series Editor

Fergus M. Clydesdale
University of Massachusetts, Amherst

SERIES PREFACE

The overall objective of the CRC Series in Contemporary Food Science is to encourage the publication of technically relevant, current information in the area of food science. As the field of food science evolves, it is becoming apparent that a broad array of scientific disciplines is essential to maximize production, processing, packaging, and marketing in order to ensure a high quality, safe, and nutritious food supply. These disciplines would include, but not be limited to food biochemistry, food chemistry, relationships between diet and disease, microbiology, physical and sensory properties, new technologies, flavor chemistry, new ingredients, food colors, the role of nutrients and their bioavailability, pesticides, toxic chemicals, regulation, food law, risk assessment and communication, and novel foods.

This will provide a broad, but nonetheless essential scope which will encompass information on those aspects of food science most critical for understanding and application in this burgeoning field.

The specific objectives of the CRC Series in Contemporary Food Science are as follows:

1. To publish advanced textbooks, professional texts, professional monographs, tutorials, and reference works, such as handbooks, for scientists at all levels involved in food science.
2. To encourage the publication of textbooks in the field of food science; in particular those books which incorporate unique and new approaches to the application of basic science to the field will be sought.
3. To identify and publish tutorials which will provide a "how-to" approach in fields of interest.
4. To publish reference material in the form of handbooks which may be utilized by scientists in academia, industry, and government.

The following topics define the scope of the books which will be considered for the CRC Series in Contemporary Food Science. These topics should be viewed as providing a general umbrella under which specific titles of in-depth manuscripts will be developed and published. This list of topics is not intended to be exclusive but merely to provide an insight into the scope of the series. The audience will include students in degree programs at both the graduate and undergraduate level, professionals in government, academia and industry, technicians, other appropriate IFT members and food and agricultural engineers.

- Food Biochemistry
- Post Harvest and Cell Biology
- Food Chemistry
- Design and Function of Macronutrient Substitutes
- Functionality of Food Ingredients

- Risk Communication
- The Role of Food Technology in Health
- Designer Foods: Evaluation and Development
- Nutrient Bioavailability
- Nutritional Consequences of New Processes
- Computer Maximization in Food Science Research
- Computer Maximization in Food Processing
- Food Product Development
- Risk Assessment
- Food Microbiology
- Toxicity Evaluation
- Pesticide Analysis and Use
- The Use of Integrated Pest Management in the Food Industry
- Techniques of Sensory Evaluation
- Quality Control
- Cellular Consideration in Food Quality
- Flavor Chemistry
- Development of Flavors
- Natural Colorants
- Novel Foods
- Food Biotechnology
- New Applications of Fermentation Technology
- The Use of Biotechnology in Producing New Ingredients
- Food Safety
- Microbiological Assessment
- New Concerns in Food Microbiology
- Physical Properties of Foods
- Food Law
- Food Regulation
- Food Analysis
- The Role of NMR in Research and Quality Control
- Water Activity

Manuscript proposals on any of the above or related topics will be considered by the Editor of the CRC Series in Contemporary Food Science. A panel of editors with unique specializations within the scope defined by the series will be utilized for in-depth review of the proposals.

Fergus M. Clydesdale
University of Massachusetts, Amherst

PREFACE

After leaving industry for private practice, I was invited to be an outside lecturer on new food product development at McGill University's Department of Food Science. This course became the framework around which this book began to take shape.

Each semester there was a critical review by students of the course and the lecturer. My reviews were always the same: the food science majors complained there was not enough science and too much marketing; the agricultural economic majors and the agri-business management majors complained that there was too much science and not enough marketing and management. Apparently the course was a success! It pleased no one except me. This is exactly what new product development is all about: steering a course through food science and technology, marketing and consumer research, and finance, where not everyone will always be happy with the outcome.

These course notes were supplemented from my own practical consulting experience, much of which arose from crisis management of other companies' new product failures. I saw why products failed and could piece together the subtle and not so subtle signs that should have suggested that all was not right during development.

There were management problems which exacerbated communication problems. I, a thousand miles away, frequently had better communication with individuals working on different aspects of the same project whose offices were literally side by side; but they belonged in different departments. I saw organizational systems that abetted these communication blockages. Technically brilliant applied scientists were pushed beyond their capabilities to communicate science to others in their companies. Non-communication resulted.

Many of the books, articles, and seminars on new food product development wrestle with only one particular aspect of product development and distort that aspect beyond recognition. Food biotechnologists do much the same thing as they promote the virtue of the many exciting developments this field will bring us, yet lose sight of biotechnology's full impact. All aspects must be put into proportion to work complementarily one with another. Then new product development has a chance of success.

Eventually, at conferences and on business, I met many of my old students. One, in particular, who while a student had been one of my severest critics, greeted me warmly at a technical conference. While we talked he apologized for his behavior and criticisms and thanked me for teaching new product development not only "the way it is" but also about the ways it could be improved.

Another student who had also been very critical of the course and of me in particular (for not arranging tours of the new product development laboratories(!) of food companies) now worked in new product development for a different division of a company that I had worked with. We mused over anecdotes from my classes. (These were mainly about classic blunders depicting how development should not be conducted.) I asked how things were at her laboratory. Her answer, "SSDD", baffled me until she explained "same stuff, different days". She proceeded to describe blunders as equally appalling as the stories I had shared in class. It appears nothing has changed much in new product development.

Large multinational companies are not at all immune from some of the errors that smaller companies with fewer resources make. The only difference is that large companies have more places to hide their errors better.

Still another student working in a contract research laboratory had not believed me about the impact of marketing research and financial considerations on new product development decisions. He has now made a full 180° turn and declared that marketing research involving knowing the consumer and the consumer's habits, understanding the marketplace, and identifying the competition plus financial projections are the most important aspects of successful new product development. It starts with the consumer.

These people and others urged me to organize my notes and experiences and publish them in a book. For their encouragement, I am thankful. This book is the result.

THE AUTHOR

Dr. Gordon W. Fuller is President, G. W. Fuller Associates Ltd., an international food consulting company based in Montreal, Quebec, Canada.

Dr. Fuller graduated in 1954 with a B.A. degree and in 1956 with an M.A. degree in food chemistry from the University of Toronto. His Ph.D. in food technology was obtained at the University of Massachusetts in 1962.

Prior to forming his own company, for 8 years Dr. Fuller was Vice President, Technical Services, Imasco Foods Ltd., Montreal, Canada, where he initiated and was responsible for corporate research and product development programs at the company's subsidiaries in both Canada and the U.S.

He has held a fellowship at the Food Research Association, Leatherhead, England, where he served as Advisory Officer and later as Supervisor of Meat Product Research.

He was an associate professor in the Department of Poultry Science at the University of Guelph, Ontario, Canada, where, in addition to teaching and research responsibilities in poultry meat and egg added-value products, he performed extension work for processors manufacturing added-value poultry meat products.

Dr. Fuller has worked in applied research and product development on tomato products for the H. J. Heinz Co. Fellowship at the Mellon Institute for Industrial Research, Pittsburgh, PA. He has worked on chocolate products for the Nestle Co., Fulton, NY and as a research chemist for the Food and Drug Directorate in Ottawa, Canada.

Dr. Fuller has lectured on agricultural economics and food technology topics in North and South America, England, and Germany. He has presented courses at McGill University in agri-business management and new food product development and currently is a guest lecturer at Concordia University.

Dr. Fuller is a member of the Institute of Food Technologists (U.S.), the Institute of Food Science and Technology (U.K.), the Canadian Institute of Food Science and Technology, and the Institute for Thermal Processing Specialists (U.S.). In addition, he has been active in the Canadian Institute of Food Science and Technology and the Institute of Food Technologists (U.S.) where he has both served on and chaired committees dealing with science and international relations.

CONTENTS

Chapter 5

Refining the Screening Procedures for the Product 125

Chapter

1

Introduction

New food product development represents a monumental investment for a company, both in money and human resources. The odds against success are disheartening. The rewards, on the other hand, can mean the continued profitability of the company.

New food products are proliferating at a remarkable rate in the North American marketplace. Estimates of how many are introduced every year vary from 2000 to 8000 new food products in the U.S. The true number of introductions per year is likely somewhere between these two figures. Kantor (1991) reports, for example, that new food product introductions in the U.S. in 1970 were 1030; they climbed to 2016 in 1980 and escalated to an astounding 9192 introductions in 1990. Friedman (1990) presents a bar graph of new food product introductions for 1964 to 1988. If one makes a simple projection of his bar chart to the year 2000, one would have to believe that there will be an astronomical number, several tens of thousands, of new food product introductions per year.

Whatever the exact figure, there is significant activity on the part of food companies to develop new products. Many in management believe new product development to be the life blood of any food company.

Estimates for the number of new food product failures are as wide-ranging as estimates for introductions. One of the reasons for this imprecision is that products may fail at the laboratory bench, in test market, or when the product fails to reach a satisfactory market share a year after product launch. At what point and on what basis was the estimate made? (Both "failure" and "success" of new food products are very misunderstood terms. They will be discussed in more detail later.)

Some estimates of the failure rate range from 1 in 6 to 1 in 20. In my own experience, gained over a 4-year period with one company's product development program, for each product that went into test market, 13 others had received some development at the laboratory bench level or had made it into the pilot plant before being rejected. Clausi (1974) estimated that in one 10-year period, the then General Foods Corporation tested conceptually, developed, and undertook home-use testing on more than seven products to find one considered suitable for test marketing. Less than half of those introduced to test markets were eventually successful. This leaves an astonishingly small number of products that will achieve their developers' goals of a market placement. The failure rate in new product development is, indeed, horrendous.

I. DEFINING NEW FOOD PRODUCTS

Before proceeding further, some definitions and explanations of the classifications used in new food product development should be introduced. In addition, some general examples of the classifications are in order.

What is a new product? There is no single definition that fits perfectly. This may explain some of the disparity in the estimates of introductions and failures. New packaging on an old product can justify its being classified as a new product. Likewise, an old, established product positioned into a new market niche is a new product, as is the introduction of an old, established product into geographically new markets, for example, export markets. A new package size of an old product is also a new product. The never-before-seen product is also a new product, though less common than the other kinds.

Each of these new product situations presents different challenges to the food company, whether seen through the eyes of marketing, manufacturing, financial, or development personnel.

A simple definition for a new product might be *a product not previously marketed or manufactured by a company*; however, this breaks down if one includes new packaging (shape or size) or if one enters a product into a new market niche — the food service sector, for example. The definition of new product development can then be broadened to *either the development and introduction of a product not previously manufactured by a company into the marketplace or the presentation of an old product into a new market not previously explored by a company*.

These definitions should not be applied too rigidly. They become a little shaky if pushed. Their shortcomings will be discussed in more depth as they arise.

II. CLASSIFICATION AND CHARACTERIZATION OF NEW FOOD PRODUCTS

New food products fall into one of the following classifications:

line extensions,
repositioned existing products,
new form of existing products,
reformulation of existing products,

new packaging of existing products,
innovative or added-value products,
creative products.

Each of these classes presents different problems in development, manufacturing, marketing, and distribution. They are discussed separately in the sections that follow.

A. Line Extensions

A line extension can best be described as a new variant of an established line of food products, i.e., one more of the same. Line extensions represent a logical extension of a family of similarly positioned products. They are food products that require:

- little time or effort for development,
- no major manufacturing changes in processing lines or major equipment purchases,
- relatively little change in marketing strategy,
- no new purchasing skills (commodity trading) or raw material sources,
- no new storage or handling techniques for either the raw ingredients or the final product. This means that regular distribution systems can be used.

There are gray areas in classifying products as line extensions. For instance, how much difference in marketing, for example, in a new product leads to that product being classified as a line extension? A ready-to-serve soup and a condensed soup are very different products in their marketing and in marketplace competition, yet there are minimal differences in their preparation. They are best not considered to be line extensions. Rather the condensed soup should be classified as a very different added-value product development.

Typical examples of line extensions might be:

- a new flavor for a line of wine coolers or for a line of flavored bottled waters,
- new varieties of a family of canned ready-to-serve soups,
- new flavors for a snack product such as potato chips,
- new flavors of a bread-crumb coating,
- a coarser or more natural peanut butter.

Development of a line extension can normally be expected to involve very little development time and, consequently, very little development money. The type of product will dictate the amount of effort required for line extensions. Additions to a line of canned bean products (chick peas, red kidney beans, white kidney beans, or pinto beans) requires little developmental effort beyond getting a reliable source of beans. Manufacturing is not disturbed by such a line extension, except for the impact upon production scheduling. Nor are production systems, such as quality control, plant maintenance, sanitation, and hygiene, affected. Minor impact on storage and warehousing may be felt.

On the other hand, in some instances difficulties may arise. For example, extending the bean line of products to an added-value product, such as a three-bean salad (some may not consider this a line extension), will present developmental problems plus manufacturing and manufacturing-support systems changes. The developers have gone from a high pH, low acid product to an acidified product. Similarly, new varieties of canned, ready-to-serve soups may necessitate extensive changes in processing lines because of the different ingredients and raw materials required.

By contrast, if a snack manufacturer extends its product line from potato chips to corn chips to corn puffs to peanuts to popcorn, these are not simple line extensions. Such products have in common only the snack food element. They may be distributed through the same channels and displayed in the same section of a retail store, but they should not be considered line extensions. Purchasing philosophies, storage facilities, and manufacturing technologies have changed too extensively.

Marketing programs are usually not affected by line extensions, but there can be some surprises. If adult flavors are introduced into a family of snack products originally positioned for children, these flavors may not be successful for either adults or children. Therefore, different promotions, advertisements, and store placements for the adult products must be considered. Similar problems might be caused by animal shapes for pasta in a sauce or children's bite-sized pieces (miniaturization) for cookies and crackers.

B. Repositioned Existing Products

A company can be very startled to find, either through their consumers' letters or through product usage surveys, that their consumers have come up with a new use for an existing product. This may allow a whole new market direction to be taken and give a new life to an existing product. The use of baking soda as a refrigerator deodorant is an example of such a repositioning. The promotion of soft drinks from the leisure market to accompaniments to main course meals, especially the breakfast market, is another repositioning that has proven worthwhile.

Oatmeal became a health food on the basis of claims made for its fiber as a dietary factor in reducing cholesterol. Oatmeal and products containing oatmeal or oatmeal fiber were repositioned. Osteoporosis as a major health problem of older women with minimal calcium intake was extensively described in magazines and newspapers. A popular antacid that contained a high amount of calcium was repositioned as a good source of this mineral.

Development time for repositioned existing products is minimal. Often, all that is necessary is for the marketing department to design and print new labels, to design a new package, and to prepare a new advertising strategy with new promotional materials. Manufacturing is unaffected. The responsibilty is with the marketing department to capitalize on the new market niche.

C. New Form of Existing Products

Putting an existing product into a new form is a radical departure from the type of new product development we have just discussed. It is also a development that is the downfall of many new product ventures. An instantized, solubilized, granulated, tableted, powdered, foamed, concentrated, spreadable, frozen, or otherwise modified version of an existing food product can involve extensive development time. It may also require major equipment purchases, both for manufacturing and for packaging. In addition, processing and the support systems for processing may be different. The company may face vastly different warehousing and distribution system problems.

The frequent fallout is that the consumer does not always appreciate the so-called improvement in the modified product. There must be a perceived advantage to the new form over the old if this is the direction new product development is to take. For example, the advantages of a dried, sprinkle-on version of a condiment sauce may not be appreciated or preferred over the traditional liquid form. Similarly, prepeeled, precut french fry style potatoes were never successful in the chilled food, retail market. However, the phenomenal successes of instant coffee, foam-dispensed dairy toppings, and spreadable margarines continue to tempt many food manufacturers into this class of development. Miniaturization, i.e., bite-sized pieces, has proven very popular with young children for snack items such as cookies and crackers.

D. Reformulation of Existing Products

The "new, improved..." product is typical of this category. Reformulation of a product to make some improvement (e.g, better color, better flavor, more fiber, less fat, greater stability) has a high probability of technical success. Usually, but not always, reformulation to obtain an improvement can be accomplished comparatively inexpensively and in a relatively short development time.

Reformulation may be necessary for any number of reasons. For example, a raw material may have become unavailable or new sources of an ingredient must be investigated. Reformulation may be needed to lower costs to meet the challenge of cheaper competitors. It could also allow a company to take advantage of ingredients with vastly improved characteristics and properties.

Reformulation may also be necessary to satisfy the consumers' demand for a healthier product, such as one with fewer calories. This could also create a new market niche for existing products. High-fiber bread, low-fat ice cream, lactose-free milk products, and baked goods free of hardened vegetable fats are good examples of such products.

E. New Packaging of Existing Products

In its simplest form, the packaging of bulk produce into unit packages typifies this category. So too does the packing of snack products, such as potato chips, corn puffs, and fruit chips, in pillow packs. New technologies such as modified atmosphere packaging (MAP) and controlled atmosphere packaging (CAP) have permitted the creation of a number of new products, providing existing products with an extended shelf life for both satisfying existing markets and allowing the opening up of new markets in a larger distribution area.

The packaging and brand-labeling of produce and meats (Gitelman, 1986) are other examples of existing products being given a new life as a new product. For example, bananas, until recently a "no-name" food, now display stickers of well-known food companies. Here, of course, development is minimal. Manufacturing involves the purchasing, inspection, grading, cleaning, trimming, storage, weighing, packaging, and distribution. However, the key responsibility in making this new product a success rests mainly with marketing a brand, the quality of which is known and respected.

A new package for an existing product may require that expensive packaging equipment be purchased. Major changes, such as from metal to glass containers, mean a redesign of the entire packaging line. Even the changeover from steel cans to aluminum cans to save weight requires an extensive overhaul of the packaging operation. Similarly, the use of plastic squeeze bottles with snap-cap lids for dispensing mustard, ketchup, or other sauces is a major packaging changeover from glass containers.

The use of the pouch and the semirigid tray (both thin profile containers) for thermally processed foods provides an added value feature and improves the quality of the product. However, the company switching to such a new style of package may have to do reformulation. In addition, extensive changes to the packaging line are required.

F. Innovative Products

The remaining two categories, innovative products and creative products, cannot properly be described without clearly defining what is meant by "innovation" and what is meant by "creation". *Webster's Ninth Collegiate Dictionary* defines "innovate" as to "make changes". *The Concise Oxford Dictionary* definition is to "make changes in". So, an innovative product is one resulting from making changes in an existing product.

By contrast, the same sources define "create" as to "bring into existence". A creative product is one newly brought into existence: the rare, never-before-seen product.

Innovative products are difficult to categorize. Generally, the more innovation (change) in a product, the longer the development time and the higher the research and development costs. The marketing of novelty may involve costly techniques because the consumer may have to be educated to the novelty. In short, the development of innovative products could be both costlier and riskier than any other path of new product development.

On the other hand, very little research and development in terms of costs and time are required for a frozen food processor to put a stew, some frozen vegetables, and a frozen pastry on a tray and call it a frozen dinner. Likewise, putting a can of tomato sauce, a package of spaghetti sauce spices, and a package of dry pasta together to make a dinner kit requires little research and development effort. Yet both were remarkably successful innovative products that have engendered many imitators. And in the nonfood category, Akio Morita, the inventor of the Walkman® (trademark of the Sony Corporation) claimed the Walkman® involved no invention and no costly research and development; it just required expert amalgamation of established inventions into a superbly marketed product (Geake and Coghlan, 1992).

New ingredients can form the basis for innovative products. Simulated crab legs, lobster chunks, shrimp, and scallops based on surimi technology have formed the basis for many seafood dishes (Johnston, 1989; Mans, 1992).

An ambitious program of innovation with all of its attendant costs for a canner of commodities, such as chick peas, navy beans, and other lentils, would be a venture into the leisure foods arena to manufacture such added-value products as hot bean dips or such ethnic dishes as hummus and hummus tahini. The demographics of many cities indicate unique marketing opportunities for the adventurous in such ethnic dishes.

A note of explanation must be given here for a term that has crept into the discussion, "added value". The late Mae West had a memorable line in the old film, *She Done Him Wrong*: "Beulah, peel me a grape!" That is added value: peeled ready-to-eat fruit is on the market shelves. The term, "added value", is used to indicate that degree of innovation that makes a product more desirable to consumers. It might be improved stability, improved functionality, better color, better texture, or better anything. Whatever it is, it is wanted by consumers. Meltzer (1991) defined "value added" (both terms, "added value" and "value added", will be found in the literature: they are synonomous) processing as "… any technique that effects a physical or chemical change in a food or any activity that adds value to a product."

G. Creative Products

Creative products are harder to define and still more difficult to present in examples. Surimi, a fish gel developed several hundred years ago, and its development into kamaboko-based products or texturized shellfish analogues would be considered a creative product. So would tofu, bean curd, and limed corn meal. Today, one might consider reformed meat products as a creative development, and certainly, extrusion to produce new puffed products is creative. However difficult they may be to give examples of, they can be characterized by the following:

- Creative products generally require extensive development time.
- Research and development costs tend to be very high.
- Marketing is costly because there can be a need to educate consumers about something new.
- Capital costs for equipment can be expensive because machinery frequently must be developed *de novo* if novel processing steps are involved.
- The introduction of creative products into a market can be very risky. Chances of failure are high.
- If the products are successful, imitators will rapidly flood the market and take advantage of the time and effort the developers of the creative product took to create and market it.

In general, the more a product is a copy-cat of an existing product, the less development time will be required. Development time may be only as long as it takes to have new labels printed. Similarly, development costs and costs of market entry will be minimal.

On the other hand, the more creative a product is, the greater these costs will be, and development time may be weeks, months, or years.

H. Summary

Figure 1 is an attempt to demonstrate three-dimensionally the difficulties accompanying the movement from known products in known markets to new and unknown markets or the movement from known products to complex innovative products as the target consumer becomes more elusive.

The traditional Y-axis is a qualitative index of increasing marketplace complexity; the farther from the origin, the more complex is the marketplace. This index of complexity includes such elements as the activity of the competition, warehousing and distribution problems, food legislation, the general economy, as well as specific economic factors in the area being serviced, the availability (or lack thereof) of marketing skills within the company, and the geography of the market, as well as the difficulty of selling novelty or of educating the consumer.

The X-axis is a qualitative index of the increasing technical complexity required for a new innovative or creative product or the difficulty of introducing into the design of the new product the necessary degree of added value demanded of the product. The farther from the origin, the more creativity, innovation, or technical complexity there is in the product. Consequently, more research and development time and costs are involved.

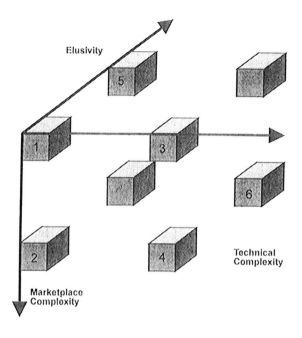

FIGURE 1. Product complexity, marketplace complexity, and consumer elusivity interactions.

The remaining axis, the traditional Z-axis, is a measure of the volatility and elusiveness, perhaps even the fickleness or incredulity (or all the preceding) of consumers. It represents the elusivity in the consumer, which is then reflected in the market. How else can one recognize the growth in "healthy" foods when there has been "... a dramatic increase in the per capita consumption of high calorie desserts, salted snack foods, and high calorie confections" (Gitelman, 1986) to the extent that the average caloric intake by consumers is up over what it was 20 years ago? Elusivity can be likened to market segmentation. If a product designed for the general public (a rare event) is redesigned for the teenage market, the consumer has become more elusive, and the market is being segmented. If this hypothetical product is redesigned again for teenagers of single parents, more elusivity is created. Marketing purists might cavil at this, claiming that this elusivity is really a variant of the Y-axis or marketplace complexity. It is not; it is not the volatility of the marketplace that is represented but the volatility of consumers.

Six numbered solids situated in Figure 1 represent typical problems faced by developers as they attempt to bring to market new products. The first solid (1), at the origin, depicts the situation of an established product in a known market (also encountered when a line extension is introduced into a known market). It is the status quo.

The second solid (2) represents the situation when the company takes the same established product into a more complex marketplace situation. For the sake of argument, this could simply be exemplified by expansion into a new geographical area; only marketing costs have increased. The targeted consumers have not changed. Development costs were not a factor in entering a new marketplace.

The third solid (3) is a product with added value (increased product complexity) introduced into a marketplace that is known to the company and aimed at consumers known to the developer. Costs now begin to escalate. Providing added value requires investment in development; introducing a new product line for which the developer is not known requires increased advertising and promotion to gain recognition.

In the situation represented by the fourth solid (4), the added-value product has been introduced into a more complex marketplace, one previously unknown to the developer. This situation could result when a product is repositioned or a new market niche has been opened for it or simply because the market area has been expanded. The targeted consumer remains the same but there is a new playing field, to use an old cliché. A popular brand of baking soda is repositioned as a refrigerator or room deodorant, to offer a satisfactory example. Researching the new market area brings increased costs because the old marketing promotions and advertising may not be suitable.

The fifth solid (5) represents another repositioning problem. An established product is targeted for an elusive consumer but in a familiar marketplace. Some examples include the popular antacid that is positioned as a calcium supplement for elderly women, or the handcream that proves to be an excellent insect repellant for campers. Costs can again increase significantly to reach these new targets, which are more elusive. The situation is risky.

The sixth solid represents the worst of all possible worlds. A technically complex product (i.e., with added value) is to be positioned for elusive consumers in a market foreign to the developer. The best example may be added-value products (cheeses made with medium chain length fatty acids, for example) for people with digestive disorders in the health care market. The risk is high; development costs are high; promotion can be difficult.

As one proceeds out along the axes, difficulties, such as delays, costs, and frustrations, increase.

Figure 2 demonstrates the hypothetical situations in Figure 1 in a somewhat more recognizable setting. Up to step four was an actual potato processing operation with which this author worked closely. Not too many years ago, only field grade potatoes were sold (see the first solid in Figure 1). They were packed in bulk as bushel, peck, or half-peck baskets or in 100-lb burlap sacks. There was neither culling, size-sorting, nor cleaning. One bought one's potatoes in the fall, stored them in the cellar, and used them up throughout the winter. This was not always convenient for many households: the smaller

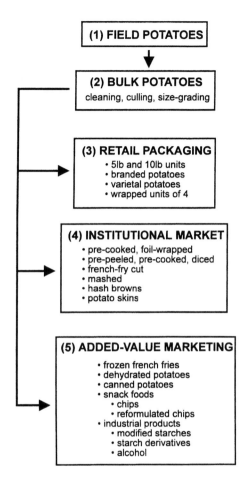

FIGURE 2. The Potato Tree, illustrating product complexity, marketplace complexity, and consumer elusivity.

family units, the increasing number of apartment dwellers, and those with changing food habits.

The first improvement (added value) was cleaning, culling, size-grading (stage 2 in Figure 2), and packaging in 5- and 10-lb units (stage 3, Figure 2). This corresponds, in Figure 1, to moving horizontally with increasing technical complexity (X-axis). Smaller unit packaging (foil- or glassine-packaged baking potatoes or other varietal potatoes) introduced new market niches and targeted elusive consumers, "live-alones", occasional potato users, or users who wanted specific varietal potatoes for a purpose (baked or barbecued potatoes). The added-value potato has moved in two directions (Figure 1): in the direction of consumer elusivity and greater marketplace complexity.

Today, many stores sell potatoes by their varietal name. They are also offered by brand names, and suggestions for the best culinary uses of that particular variety are described. A subtle complexity has now been introduced. Previously, the processor seemed to be distanced from consumers; now branding has come to the fore. Processors and consumers have been brought together.

The next logical and simple processing step to provide added value was to prepeel, dice, and slice the potatoes. This added convenience by minimizing preparation time and concentrating waste in a central location. It also opened up a new market, the institutional market (Figure 2, stage 4). Food service demanded convenience and variety in the form of new products.

But new problems were introduced: greater technology was required to reduce losses due to spoilage with a more fragile product and to prevent hazards of public health significance. That is, more technology was needed to provide the greater added value. In terms of Figure 1, one might accept that the third solid represents the situation. A new market meant new marketing techniques. A distribution network suitable for a fragile product had to be established. Now it might be argued that the fourth solid in Figure 1 more accurately describes the situation.

Further sophisticated processing serves the needs of the food service sector with preprepared hash browns, baked potatoes, partially cooked french fry style chips, stuffed potatoes, and so on. Each convenience introduces the need for greater technology.

A divergent path of development (step 5 of Figure 2) brought the processor to the canning, freezing, and dehydrating of various potato products. Consumers now have the convenience of having mashed potatoes from a package or french fries without the mess of a deep fat fryer.

The leisure food market — fun foods, food for snackers or grazers, food for people on the run, finger food — is a phenomenally successful market. Still further processing took the potato processor into thick cut or thin cut, crinkle cut, skins on or off, flavored or plain, extra crispy or reduced-fat chips, or chips reformulated into new shapes.

Here, in Figure 2, the whole gamut of new product types can be seen from the creative, reformulated chips to simple line extensions (branded potatoes and varietal potatoes). It also demonstrates the problems that arise as development requires more technology and more costs and as new markets are opened up. As one progresses down Figure 2, there is an increasing technical complexity of the products. Marketplaces for the products become more complex, demanding a greater marketing capability to cope with more elusive targeted consumers.

The analogy with Figure 1 can be continued, albeit with the example of potatoes making it somewhat difficult to push. Industrial products certainly drive technical complexity far to the right and, at the same time, push market complexity into new market areas, for example, that of food ingredients. Reformulated chips and other products demonstrate increased technical complexity (solid number 3), represent a leisure snack market (solid number 4), but also appeal to a more elusive consumer, i.e., the up-scale, older consumer (solid number 6).

III. WHY GO INTO NEW FOOD PRODUCT DEVELOPMENT?

If new food product development is fraught with so much difficulty, if it is so costly, and if it has a high rate of failure, why go into it? Would it not be simpler to coast along with the existing products?

This certainly would be simpler but it would not be profitable for very long. Food companies must grow to make money and to survive. As John Maynard Keynes put it so succinctly in *A Treatise on Money*, "The engine which drives Enterprise is not Thrift, but Profit." New food products are the major avenues open to a food company to be profitable and to survive. Some marketers would argue that new food product development is the only path the food company can follow for survival.

The need for new food product development can be seen to be driven by five dominant forces:

- All products have life cycles. That is, they enter the marketplace, flourish for an indeterminate time, then die, and must be replaced.
- A company's management may adopt a policy that requires an aggressive growth program to satisfy long-range business goals.
- The marketplace may change, requiring new products more suited to respond to the changes.
- New technology may make new food products available and new knowledge may tailor new food products more suited to the lifestyles of today's consumers.
- Changes in government legislation, health programs, agricultural policy, or agricultural support programs may dictate that development of new food products be pursued.

A. Product Life Cycles

Each product has a life cycle, depicted stylistically in Figure 3a. The vertical axis in Figure 3a is an index of a product's acceptance. This could be either volume of cases sold or sales dollars. Here, case volume of product sold was chosen to follow life cycle. Volume is obviously low when the product is first introduced. Volume grows as consumers buy, are satisfied, and make repeat purchases. In time, the market becomes saturated; the volume levels off and gradually declines as consumers switch to other products.

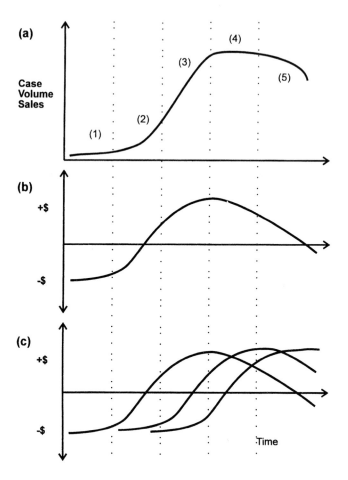

FIGURE 3. Product life cycles and profitability. (a) Typical product life cycle; (b) the profit picture; (c) the contribution of new products to profitability.

Five distinct phases to the life cycle (Figure 3a) can be discerned:

1. the introductory period, heavily supported by promotions, in-store demonstrations, and couponing;
2. a strong growth period, when first-time consumers are repeat-buying and more new consumers are being attracted. There is positive growth in sales. Growth continues as new markets are opened, but promotion and expansion are costly;
3. a beginning decline in the growth of the sales volume. Growth is accelerating negatively;
4. a period of stability, better described as a no-growth period of a stagnating market;

5. a beginning decline in volume. New competitive products adversely affect sales. Promotions prove too costly to maintain sales volume.

Life cycle curves can be generated for product categories as well as for specific products within a category. Instant coffee, as a product category, could be described as being still in the growth phase. Nevertheless, the leading brands of instant coffee have changed places as each goes through different stages of the cycle at any given time. The sale of flour, another product category, had for years been in a no-growth phase that was only slightly ruffled by the advent of cake mixes; now it is enjoying a modest growth as many households are returning to the art of home baking. During a period in the 1970s when meat prices soared because of a scarcity of beef, the sale of meat substitutes and extenders grew dramatically and then plummeted drastically when meat became plentiful and prices fell. Life cycle curves are as varied as products are.

More revealing of the success of a product than sales is the profit brought in by those sales.

The introductory phases will have minimal net profit (see Figure 3b, phases 1 and 2). This period bears the costs of research and development, as well as carrying the additional heavy costs for promotion and market penetration. There are improved profits during the growth phase (phase 2, Figure 3b). The improvement continues in phase 3 but toward the end of this phase begins to drop off as costs for market expansion and costs to support the product against the competition begin to take their toll. During the no-growth phase (phase 4), costs to support the product match the profits but rapidly drop off. The company eventually sees the product as an unprofitable item that cannot be maintained. Manufacture of the product ceases.

To maintain the viability of the company, there must be replacement products ready to be launched to continue the flow of profits. In Figure 3c, two additional products have been launched by the company. These products maintain the company's net profit picture in a healthier state. The company does not wait for its profit to dip as its product loses ground in the marketplace. It keeps a series of new products in various stages of development. Bogaty (1974) claims that for every one product on the national market, two should be in test marketing. For each of those two, there should be four in the last stages of consumer testing. He works this pattern back until reaching the level at which 32 product ideas are being screened for development consideration. The cumulative profitability of these new food products, if they are successful, assures a good return on investment.

B. Corporate Reasons for New Product Development

A company exists to grow and to be profitable. Senior management, under the direction of the owners or the shareholders, follows a corporate business plan that will set out specific financial and growth objectives.

Food companies can achieve growth only in a limited number of ways. For example:

1. Expanding into new geographic markets is a possibility. It can be expensive. For products with short shelf-lives, the distribution system and its costs may limit expansion. Export markets present their own unique hazards.
2. Trying to achieve market penetration and a greater market share of existing markets dominated by the competition is a means of growth. This method also means slugging it out toe-to-toe with competitors. Large sums of money are required for advertising and promotion.
3. New products can contribute to growth and profitability (see Figure 3c). Such products must be developed to bring in new profits. There are, however, the costs of their development to be considered.

Each of these avenues to growth comes with its own costs and associated problems. The first two avenues to growth can be very expensive. New product development, with all its hazards, strengthens the company's base and brings in profits.

Complementing these measures to achieve growth is the need to reduce expenses and overhead costs. This can be accomplished by reduction of staff, implementation of an energy conservation program, improvement of processing efficiency, adoption of a waste management program, and adoption of a sound process and quality control program to reduce losses through overfill, rework, and product returns. These thrift measures may help the company's profitability but are of limited value for growth. Profit drives the enterprise, not thrift, to paraphrase Keynes.

When plants operate seasonally or have a slack season, there is an incentive to spread production evenly throughout the year. This keeps trained workers employed throughout the year, reduces plant overheads, provides a more steady cash flow, and benefits the community. A plant operating year-round is more profitable than one idle most of the year. The off season can be used to produce new products, putting the underutilized plant to work.

C. Marketplace Reasons for New Product Development

The marketplace is constantly changing: it is not a static organism but a dynamic one. Several factors contribute to this volatility.

Food purveying is taking many new forms. Warehouse stores, mail catalogue shopping, and teleshopping are changing consumers' buying habits. The traditional supermarket is adopting a new concept in retailing: it is becoming a collection of food boutiques. Small mom-and-pop stores are becoming 24-hour convenience stores. The abundance of restaurants, diners, take-out food outlets, and deli counters in supermarkets makes prepared ready-to-eat food abundantly available.

The "consumer" is a very volatile influence in the various marketplaces. But who is this "consumer"? The "consumer" may be the daily shopper or the more traditional, once-a-week shopper in the supermarket, a restaurant-goer, a fast-food chain buying its pickles and relishes in the food service arena, or the government buying in the institutional marketplace for the military. As both the consumer (whoever or whatever that may be) and the marketplace (wherever or whatever that may be) change, food manufacturers serving that market sector must respond to that change quickly.

Another factor in the marketplace influencing its dynamics is the competition. The launch by a competitor of an improved product requires some reaction from companies with products whose sales may suffer from this introduction. A new food product launch requires some retaliation on the part of other companies with similar products. This retaliatory action may involve new pricing strategies, promotional gimmicks, or the development of new products to combat the competitor's intrusions.

The profile of consumers in any marketplace, but in particular in the retail marketplace, is constantly altering. The result is changing consumer buying habits. There are many factors causing this. Population movements bring changes in the ethnic background of neighborhoods. With these movements come consumers with different food needs. Populations age. Any geographic market area is in a constant state of flux with respect to the consumers in it, their ethnic makeup, their incomes, their education, and their lifestyles. Empty downtown city neighborhoods enjoy a rebirth as fashionable areas for young professionals.

This fluidity in the marketplace must be accepted by the food processor as a challenge. No single product can be expected to answer all the demands of consumers. Only a battery of new products will suffice. New products are needed to satisfy emerging market niches. The subtle and not so subtle changes in the marketplace can be a great motivator for product development.

D. Technological Reasons for New Product Development

Science is continually providing increasing knowledge about the physical world, human health, the environment, and materials. This knowledge is then translated into processing technologies and products that assist us in our daily lives. These advances have led to the realization of products and services only dreamed of before.

With computers linked via telecommunications to information databases, any company can access vast quantities of business and technical information that previously were accessible only in specialized libraries in major urban centers. Expert technical information to assist in development programs is presently available to even the smallest food processing company in remote farming regions.

One can find similar examples of technical advances in food packaging. Once, the mainstay of packaging was the three-piece can with its lead-soldered side seam. This gave way in favor of the two-piece seamless can. A multitude of materials are now available, where at one time the packaging industry relied primarily on tin-coated steel, glass, and aluminum. Plastics, coated paperboard, composites of aluminum, plastic, and paper, and even edible food cartons have permitted the manufacture of a wide range of containers with unique properties to preserve and protect the high quality shelf-life of foods. Containers are now microwaveable, edible, degradable, and recyclable.

Greater knowledge of food spoilage mechanisms and preservative technologies has given rise to new products with better stability, quality, and nutrition. Improvements in retorting technology and equipment have elevated "canning", the workhorse of food preservation, to a technique that gives added value through improved texture, color, and flavor. New ingredient technologies can modify flavor, texture, and other attributes of products and maintain these through strenuous processing procedures.

Advances in nutritional knowledge have produced a growing awareness of the role of food to health and specifically of the relationship of certain foods to disease. This gives every indication of spawning a new class of preventive foods: to prevent cancer, heart disease, aging, etc.

Food products are now recognized as much for the absence of something (sodium, cholesterol, saturated fats, refined sugars) as for the presence of something (calcium, fiber, mono- or polyunsaturated fats, antioxidant vitamins).

Technology can change the marketplace in many ways. For example:

- Consumers become more knowledgeable about foods and nutrition and become more discriminating shoppers.
- Food manufacturers become adept at developing new products for the marketplace.
- Better techniques to research consumer behavior provide marketing with improved skills in planning products for the consumer.
- Retailing uses its knowledge of consumer behavior to attract consumers (the use of odors, for example) and to service their needs.

A food processor cannot afford to be unaware of developments in science and technology and the impact these advances have on consumers and food retailers.

E. Governmental Influences on New Product Development

The objectives of government, with its promulgation of food legislation, are (Wood, 1985):

1. to ensure that the food supply is safe and free from contamination within the limits of available knowledge and at a cost affordable by the consumer;
2. to develop with food manufacturers standards of composition for foods, as well as labeling standards;
3. to maintain fair trading and competition among retailers and manufacturers in such a way as to benefit consumers.

The first objective certainly has an impact on the technical development of food products, as does the second. Both the second and third objectives will influence marketing personnel as they prepare labels and promotional material.

There is government at the federal level, the state or provincial level, and the municipal, local, or county level. Government at all levels strongly influences the business activities of food companies.

Compounding the influence of these official levels of government are two more: international bodies and quasigovernmental agencies. These also can bring regulations to food manufacturing, to international trade specifically, and to new food product development indirectly. Some examples of regulatory bodies at the international level are

- General Agreement on Trade and Tariffs (GATT)
- European Economic Community (EEC)
- International Standards Organization (ISO) and its recently published ISO 9000 series on quality control and management
- Codex Alimentarius Commission under the joint direction of the FAO/WHO Food Standards Programme
- the many bi- and tri-lateral trade agreements that are arising to counter the EEC

Quasigovernmental bodies do not have the legislative powers of the various tiers of government but do have the support of government or the effect of law. These bodies can establish regulations that participating parties must adhere to. Classic examples are the various marketing boards that exist in many countries to regulate the local supply, importation, and price of many food commodities and ingredients derived from them.

Other examples are professional and trade associations. These may establish rules of conduct and wage scales for their members.

Government, international organizations, and quasigovernmental bodies affect a wide assortment of activities through various means. Some of these are tabulated in Table 1.

A few of these require further discussion. Governments often provide attractive opportunities for development in order to stimulate depressed regions of their countries, stimulate depressed industries, or utilize specific

TABLE 1. Various Food Business Activities Over Which Governments in Different Forms and at Different Levels Exert Influence

Activity	Influence
Fiscal policy	interest rates for development loans
	grants-in-aid; research funding
	taxation policy
Patents and copyrights	copyright protection and licensing
	research funding if guarantee of patent protection
Trade barriers	tariffs and protectionism
	standards of food product identity
	availability and cost of ingredients
Environment protection	waste disposal
	recycling or reuse requirements for packaging materials
	energy utilization and disposal
Marketing and trade	product or advertising claims
practices	billboard and advertising placements
	zoning bylaws
	store hours
	container sizes
Employment practices	OSHA and worker safety
	unemployment benefits
	minimum wage levels
Health policy	nutritional guidelines
	nutritional labeling
Agricultural policy	support programs for commodities
	availability of commodities
Consumer protection	product safety; safety of agricultural chemicals
	labeling, product names, comparative advertising
	inspection services

commodities. These opportunities allow eligible companies to undertake development programs, to modernize equipment, or to utilize new technologies for product development.

In developing national health policies, governments establish nutritional guidelines. Food companies may want to adopt these guidelines as a company policy for all their products in order to promote the nutrient value of their products. They may have to reformulate products to meet the guidelines. Where standards of identity or nutrient guidelines for food products differ from one country to another, a food company with a strong export trade may need many reformulations of a product to meet the requirements of other jurisdictions.

Governments also regulate consumer protection and product safety. Several years ago, the U.S. government's intervention in the banning of saccharin and cyclamates caused many manufacturers of dietetic foods and low calorie soft drinks to reformulate their products. Some companies chose to find suitable noncaloric alternatives to reformulate with. Other companies developed canned fruit packed in water or packed in their own fruit juices.

Government, with the legislation it enacts, in the policies it adopts, and through the trade alliances it makes, can have a tremendous influence on new product development.

IV. PHASES IN NEW FOOD PRODUCT DEVELOPMENT

Most authors divide new food product development into several distinct phases. Very few agree on the numbers, the order, or the names of the phases. This is not a problem in analyzing the process of new food product development. However, when developers interpret the phases as a sequence, a one-after-the-other cascade from ideas through to a final finished product, they misinterpret the process. The phases do not start, proceed, and then finish, with the next phase then beginning. The phases are not, strictly speaking, sequential: they often overlap and are concurrent. Projects might even return to the conceptual phase for a complete rethinking of the concept statements as new information arises.

The starting point most generally agreed upon is to establish company objectives and identify consumer needs (Figure 4). Next comes the generation of ideas for new products whose successful development would then meet these company objectives and satisfy these perceived needs of targeted consumers. The thick, solid lines in Figure 4 show the advance of product ideas and later the product itself. Thin solid lines indicate flow of data and information.

The next phase is to winnow all the ideas and to reduce their number to a manageable few deemed to be the most worthy. There are three parallel screening criteria that are used:

1. Is the idea feasible within the time frame required by the marketing department? This answer should be provided by the manufacturing, engineering, and research and development departments.
2. Does the idea meet a perceived need of the consumer? Marketing and consumer research will determine this.
3. Will a financially sound business plan based on these new products stand up to critical analysis?

At the next phase, the technical skills of the research and development department are brought into play. They develop bench-top prototypes that match the product statement as closely as possible. The next few boxes in Figure 4 might suggest that research and development are dominant. This is incorrect. As development proceeds through various product stages, information on raw materials, ingredients, packaging materials, and equipment requirements for

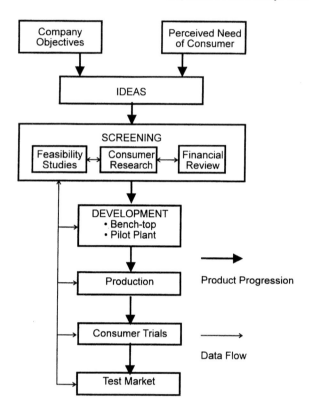

FIGURE 4. Phases in new food product development.

costing, as well as information on the acceptability of prototype products, are constantly being fed back into screening. Refinements based on the screening evaluations are fed back to the technical development group.

A series of parallel events begins (not depicted in Figure 4), based on the data obtained throughout the development process. An analysis of the business plan with more complete information on ingredient, processing, and marketing costs can now be refined by the financial department. Ingredients and packaging materials can be sourced. Marketing people can draft labels and label statements, refine consumer analysis, plan some marketing strategy and develop promotional literature. The manufacturing department can determine what production facilities and manpower may be required.

As data produced at each phase is transformed into useful information, drastic decisions must be reached. Go/no-go decisions must be made. If the concept must be changed, this will require an alteration in the product statement. Changes in the direction of product development will occur as a result. It may be necessary to "return to the drawing board" many times. Development is a constantly evolving process.

By the time production samples have undergone consumer trials, the company should be able to decide whether to go into a test market. Marketing, depending on the wishes of the company, may conduct mini-market tests, run market tests in only one or two cities, or go directly into a regional launch (Figure 4).

The final phase of any new product development is an evaluation of that test. If the test market was unsuccessful, then the weaknesses must be learned and corrected before the next new product is developed. If it was successful, then why? The strong points of the process must be recognized, which will allow the company to capitalize on these strengths in future development programs.

Chapter

2

The Generation of New Product Ideas

A shout of "Eureka!" followed by the appearance of the scantily clad body of Archimedes heralded the discovery of a basic physical principle that has found uses in many branches of science. The idea, so we are told, came as he entered his bath and noted the displacement of the bath water. Would that all ideas came so spontaneously! It makes an interesting apocryphal story. If truth be told, the basic concept was probably the result of a vast amount of thought and experimentation, which only clicked together at the last moment.

Creative people — writers, artists, and inventors — seem gifted and inspired to the ordinary people of this world. What is not fully appreciated by the rest of us is that creative people are also very hard workers, who fill many a waste paper receptacle with discarded ideas, outlines, and schematics. Ideas come only after much study, thought, research, and experimentation, plus plain hard work.

Generating new product ideas within a food company is not easy. Those responsible for new product development see ideas that they think the company should pursue from their own unique perspective, be it technology, marketing, production, or finance. They are inclined to view the development process through blinders, seeing only a narrow field dominated by their background. That is, the shackles of training, education, and experience blind each to new ideas other than one's own.

For example, technical people have a need to satisfy the professional demands of the particular scientific disciplines they serve. They see ideas as opportunities for technological challenge. Often they fail to understand that ideas must lead to products that satisfy the demands of consumers, on whom the company depends for its survival. They frequently consider the ideas of others as impractical. In particular, ideas from sales and marketing personnel within the company are considered not only impractical but — sin of sins! — unscientific.

On the other hand, marketing people live in an imaginative world where hyperbole rather than understatement dominates. They brand technical people and their ideas as unimaginative and too negative. According to marketing personnel, technical people have tunnel vision created by their scientific disciplines.

Manufacturing personnel, meanwhile, look at ideas from both technologists and sales and marketing personnel with horror and wonder how much their production schedules will be disrupted or what new production equipment will be required. The ideas of this department are sound and practical and, incidentally, will cause the least disruption of their beloved orderly routines. They, after all, make the products that make the money.

All the while, financial people within the company view with disdain all ideas that require spending more money on uncertain projects. They have their own ideas of how to make money for the company in the financial markets with which they are familiar.

This description of attitudes may seem harsh, but it is not far off the mark for many companies. I sat in on a food product development meeting attended by senior management, food scientists, and marketing and production personnel to discuss various product ideas. That meeting was demolished when the senior vice president of finance abruptly pointed out to the assembled staff that by transferring company funds into various foreign currencies or by investing in bonds or stocks the financial department would make more profit for the company more surely than they would.

This, then, is the world into which a new food product development manager may be thrown. Instead of an atmosphere of creativity or innovation, an atmosphere of bitterness, enmity, and rivalries may result. Such divisiveness in any company is not conducive to the generation of ideas for growth through new product development. There must be an atmosphere wherein creative and innovative thinking can generate ideas. Product managers, whether they be presidents of small food companies or more junior product managers of large companies, must learn to encourage individual personnel to perform as part of a team, with each element cognizant and respectful of the strengths of the other members.

I. ENCOURAGING AN ENVIRONMENT FOR GENERATING IDEAS

Edwin Land (1963), then President and Director of the Polaroid Corporation, on the occasion of the 50th Anniversary of the Mellon Institute, said:

> In our laboratories we have again and again deliberately taken people without scientific training, taken people from the production line, put them into research situations in association with competent research people, and just let them be apprentices. What we find is an amazing thing. ... In about two years we find that these people, unless they are sick or somehow unhealthy, have

become an almost Pygmalion problem; they have become creative. If there is anything unpleasant to an unprepared administrator it is to find himself surrounded by creative people, and when the creative people are not trained it is even worse. They have two unpleasant characteristics: first, they want to do something by themselves and they have some pretty good ideas that do not fit in with policy; secondly, they have the most naive, uncharming and unbecoming direct insight into what is fallacious in what you are doing, and that, of course, is a blow to policy. I do not want to romanticize these people. I am simply reporting on what we seem to find is a fact ... and you have to find out what to do with these awakened people.

(Managing creativity can be a problem; this shall be treated in a later chapter.)

Establishing an environment in which people can be creative and generate new product ideas is a top priority. Land (1963) latched onto the idea by trying to develop such an atmosphere for creativity and innovation. Too often, one's own disciplines fetter one's mind with unwritten or even written strictures on what is the proper, accepted, or correct way to do something. A set of laws, regulations, orthodoxy, or peer pressure govern one's thinking and funnel it along acceptable (proper, orthodox, correct) lines. This direction of thought can stifle creative and innovative approaches to problems. Land (1963) took people out of the humdrum constrictions and allowed them to develop and become creative.

Children are unencumbered by this rigidity of thinking. It is only acquired as they grow up, become educated, encounter peer pressure, and begin to fear looking the fool or being on the outside. Young children have a wonderful capacity to put together unrelated ideas in implausible and improbable ways, as can be seen in their stories or drawings. There is no embarrassment in freely associating seemingly bizarre ideas. The undisciplined and childlike mind has not yet been confined to the path of "correct" thinking. This spirit of free association of ideas was very apparent when I worked as a YMCA instructor and youth leader. I played a story game with my young charges. I started the story, reached an impossible situation with the story characters, and then passed the story on to one of the children. The rules were simple: the next storyteller had to start after a count of five; no magic was permitted and no violence; new characters could be introduced at any (and usually most appropriate) time. Children were eliminated when they could not either extricate the characters from whatever misadventure they were in or start the story on time. Few children were ever eliminated, but they loved to eliminate, it grieves me to say, their leader. The inventiveness of these children was mind-boggling.

James (1890) describes genius as "... little more than the faculty of perceiving in an unhabitual way." Another writer (Anon., 1988a) commented as follows about creative people:

Highly creative people are eccentric in the literal sense of the word. They have less respect for precedent and more willingness to take risks than others.

They are less likely to be motivated by money or career advancement than by the inner satisfaction of hatching and carrying out ideas. In conventional corporate circles, such traits can look quite eccentric indeed.

Stuller (1982) recognizes six types of creativity. These are, with some examples:

- theoretical (Albert Einstein, Sir Isaac Newton)
- applied (the Curies, Henry Ford, Bell, and Edison)
- inspired (artists and composers)
- imaginative (writers and poets)
- prescriptive (thinkers such as Plato, Machiavelli, and Martin Luther)
- natural (dancers, musicians, singers, and sports figures)

However he classifies creativity, he recognizes a common theme throughout: the protagonists have the ability to make associations of dissimilar things. That is, their creative ability is not compartmentalized. Like children, creative people have no fear of freely associating quite dissimilar ideas to emerge with entirely novel approaches to problems.

At the beginning, there must, then, be an atmosphere for idea generation where strictures imposed by discipline, training, peer pressure, and peer ridicule are removed. An appeal must be made to the "childlike" quality in people. In this atmosphere, the purpose is to glean ideas. This means that *all* ideas deserve a respectful hearing. All sources of ideas must be considered, whether they emanate from the boss's wife (where many of them do come from in both small and large food companies), the janitor, the technical director, or the salesman restocking shelves on his delivery route.

There can be no room for negativism. Comments such as "We did that 20 years ago" or "What good or use is that?" are clearly not going to promote an atmosphere in which creativity or innovation will flourish. People will hold back their ideas fearing a rebuke — peer pressure — from their associates. The NIH (not invented here) syndrome must not be allowed to prevail in the creative atmosphere where a company wants to generate new ideas. Discouragement of ideas is taboo. Later, all ideas can be screened, evaluated, and accepted for further development or rejected for valid, documented reasons.

Slavish attention to facts, logic, or reason (the refuge of technologists) will stifle, at this early stage, any ideas leading to creativity and innovation. According to Sinki (1986), such technical snobbism rates high in creating a technical myopia. This he defines variously as the inability to make crucial connections between ideas and applications, the difficulty in "making the translation from abstract to concrete terms"; or the reason "why ideas get aborted in their early stages". This is the blindfold that one's training, education, and experience can put on the free association of ideas from other disciplines to create something greater. Of course, information is important; of course, it can assist the generation of ideas; but it can also limit the capacity for the "bouncing of ideas off people". Too much information can intimidate and funnel thinking into accepted but uncreative channels.

There must be good communication between people from the various disciplines within a company. As a factor in technical myopia, whereby there is the separation of theory and practice, Sinki (1986) cites a lack of communication between scientists and entrepreneurs, i.e., those who make the idea work. Lack of correlation between unrelated disciplines, another indication of broken communication, contributes heavily to technical myopia.

The final contributors to Sinki's depiction of technical myopia are lack of perseverance and a failure phobia, that is, not "having guts" to take ideas on to innovative products. How much of creativity or innovation is plain, old fashioned hard work?

II. SOURCES OF NEW FOOD PRODUCT DEVELOPMENT IDEAS

Ideas for food products are everywhere. An assignment given to my students for a course on new food product development required that they keep a diary of all food purchased and eaten during a 1-week period. Portions purchased, cost per serving, type of packaging, country of origin, whether it was an added-value product, and whether special handling, storage, or preparation was required were noted. After the purchases were discussed in class, the students were asked to evaluate them with respect to whether their needs as students were satisfied. Then, they were to determine what they would have preferred to have eaten that would have fit in better with their lifestyle as students; what would have made life, their "food life", more enjoyable.

This exercise produced a wellspring of ideas that would have delighted any product development manager. When confronted with a what-would-I-prefer situation, they began to look at what constituted their lifestyle: students are cash poor, have long working hours, have inadequate or nonexistent food preparation facilities, and endure academic pressures. They came up with ideas fitting their needs. Such an exercise should be required training for all managers of new product development.

Some of the pressures that drive a food company into new food product development discussed in the previous chapter can be sources of ideas for products. A new technology, such as extrusion technology, opens up a wide vista of ideas for snacks. Greater understanding of the principles of water activity and hurdle technology (see later) have given rise to new categories of semimoist foods, ranging from dog food to fruit leathers to chilled foods. The need to reduce costs by upgrading a useful byproduct into an added-value product may motivate a company into doing research to find new uses for the byproduct.

However, it must be remembered that new product development is neither an exercise to use up waste product nor a means to display a company's technical skill for all the consuming world to see. Technical innovation, per se, will never sell a product, except to other technocrats. Nor is new food product development meant, necessarily, to be novel for the sake of novelty. New food products are meant to fill perceived needs of consumers and to meet the expectations of those consumers for those products.

A. Internal Sources of New Ideas

There are many areas within both large and small food companies from which ideas for new products can spring or can show the way to new market opportunities for new products. These sources need to be researched.

1. Sales Representatives

First and foremost as sources of ideas for new products are the company's own sales force. They have closest contact with both the consumer, as they restock store shelves, and the retailer, as they discuss orders. These people meet consumers; they see competitors' products. Their discussions with store management can reveal weaknesses in products, packaging, or deliveries. Store returns may point to problems with formulations, and complaints by customers to store management may uncover faulty preparation instructions.

The sales force is the closest resource a company has to what is happening in the marketplace. They provide the earliest signal, that should be heeded by the company, of activities in the marketplace. By listening to retailers and consumers, sales representatives present their company with ideas of what consumers want in a new product and information about how much consumers are willing to pay for this innovation. Some of the product's design work has been done for the company who listens. Salespersons should be seen as sensors, the eyes and ears of the company.

A novel approach was undertaken by one company to give this consumer- and marketplace-awareness to its engineering, research and development, and distribution departments. Members of these departments were given opportunities to go along with a sales representative on customer contact visits. Such visits served two purposes. Customers were impressed by the concern displayed by the company. Staff of these sometimes insular departments were exposed to the environment of consumers and the marketplace where their products competed with others.

2. Consumer Correspondence and Communication

When consumers take time to write letters to or call companies, those people are expressing a need to be heard. They may want to vent anger about a failed product, express pleasure about a favorite, inquire about a product's suitability for some diet, seek clarification about cooking instructions, or offer some useful information that they found when using the product. Whatever the need is, consumers have something important to say.

Many food companies now put hotline or 1-800 telephone numbers on their labels so that consumers can telephone directly to the consumer relations department. Not only can consumers' needs be addressed immediately, but companies, by using staff skilled in consumer relations, can elicit background information. Valuable psychographic information so obtained assists food companies in their marketing strategies and new product development plans.

Consumers, whether they are complaining, inquiring, or suggesting, should be listened to. When the communication is a complaint, the company is looking at the tip of an iceberg. More is hidden beneath the water than shows above. If one complaint is received from a consumer, how many consumers observed the same defect but did not contact the company?

Estimates to evaluate the significance of complaints vary widely. Ross (1980) estimates that for each consumer who complains, there are eight who did not but will not try that product again. A less conservative estimate indicates that 20 or more consumers do not complain for each consumer who does. Graham (1990) reports that for each consumer who complains, there are 50 who do not.

Complaints first should be acknowledged. Next, they must be thoroughly investigated to determine whether something serious happened in manufacturing or is happening in the distribution channels. Finally, the complaints should be classified according to their nature, the product involved, and its code identification, identification and location of the complainant, where the product was purchased, time of year of the complaint, and nature of the retail outlet, along with as much other product identification as can be obtained.

Careful analysis of the data may reveal a hidden product defect (a common recurring complaint), instability (seasonal complaint), or a package weakness (geography pinpointing a problem in distribution), to state only some obvious conclusions.

From these records, astute, consumer-oriented companies will see opportunities for new products. They may see a need to improve a food product by reformulation or by new process technology. Or repackaging may provide consumers with higher quality, more stable products. If products consistently fail to meet consumers' needs as evidenced by complaints, then companies may come up with ideas for new products more closely designed to meet these needs. Daniel (1984) describes a very elaborate computerized system for organizing consumer complaints of a hypothetical food company with a 1-800 telephone number. The system is designed to serve a quality control function but could be adapted for an approach more oriented to consumer relations. Cooper (1990) discusses the development of a computerized consumer complaint system to be used within the consumer services of a large food manufacturing company.

Complaint files are a valuable source of ideas; they should not be handled facilely by marketing personnel who might send a form letter with some coupons and thereafter discard the correspondence. All consumer complaints, indeed any consumer communication, should be catalogued, not merely filed, and cross-referenced for the nature of the complaint, for suggestions for a novel use of the product, for opportunities for improvements, or for ideas for new products put forward by consumers.

3. Internal Product and Process Research and Development

Experimental processing trials or experimental packs are carried out routinely in food plants for a wide variety of reasons. Most companies document such experimental trials on processes and products. The reports are kept on file.

Researching these records for ideas can be a useful exercise. While much may be historical or have limited value because of later improvements in equipment and ingredients, there can be clues to new product ideas. At the very least, there will be records of ideas or products that were rejected before. Reasons for rejection of products or processes 10, 5, or even 2 years ago may not be valid today. Projects impossible a short time ago may, through advances in process, package, and ingredient technology, be within present processing skills.

Very few food companies are today without computers and computer software to organize and catalogue all the reports of process and product experimentation that have been conducted within their walls. Such an information management system is invaluable for the company interested in a strong program of research and development.

Unfortunately, in many companies, experimental plant trials are either not written up at all or are poorly documented. Filing is done not in one central repository, but in privately held files scattered throughout the plant. This lost information represents a significant loss of research and development dollars. A new research team will be doomed to waste time and money "reinventing the wheel" when they could have been building on the work of past groups. If a test is worth doing, it is also worth recording and filing the results competently so that others can find the data and understand the information contained in the documents.

4. Collective Memory

If filing in-house data for computerized retrieval, as presented in the previous section, is pushed further, the company can develop a collective memory of what happened "before" and how problems were solved in the past. In the history of any company, people have encountered and solved problems in manufacturing; they have overcome short supplies of raw materials by combinations of reformulation or novel processing; they have resolved failures in products or packaging. Ideas and experiences that company employees have had are important to the company as a resource. If the knowledge and experience of the cadre of old-timers and pensioners can be organized into an accessible body of catalogued information, then companies have a valuable asset. Whitney (1989) describes the development of expert systems using the skills and experience of company personnel.

This collective memory is a tool that no company should allow to be lost. Everything should be recorded and referenced: all data and information from all sources, especially from the old-timers, are valuable for the clues and ideas hidden in them. A general discussion of expert systems is provided by McLellan (1989).

B. External Sources of New Ideas

Ideas generated exclusively from within companies carry risks. They may bring an inward, too introspective range of ideas for development. Exploration of opportunities derived from ideas from outside the company will balance this.

1. Food Conferences, Exhibits, Trade Shows, and Research Symposia

Attendance at foreign or domestic food exhibits and trade shows allows food personnel access to the widest arrays of products and technologies pertaining to foods and ingredients. Attendees can see and taste a wide variety of products from around the world. They can discuss products and the markets the products serve with the manufacturers' sales representatives. They have opportunities to see new ingredients and technologies demonstrated. They can discuss their problems with technical sales representatives. In other words, they can have maximum exposure to a broad range of products from many countries that can be sources of new product ideas. Awareness of food products in other sections of the country and in other countries provides a company with enough lead-time to act with new ideas rather than to only react to changes in the marketplace with hastily and poorly conceived ideas.

Similarly, foreign or domestic technical meetings on food topics are sources of ideas for alert, prepared attendees. They can hear technical papers, question authors about recent developments, and make useful contacts with experts in such varied fields of technology as flavor encapsulation, water activity, or irradiation, to mention only a few. Information and contacts can be very useful at some later date in product development.

Who should go to shows and conferences? Those who will return the most benefit from the experience to their companies is the obvious answer. These will be those most intimately involved in the development process, that is, marketing, manufacturing, and technical personnel.

Unfortunately, there is a real world. In this real world, it is regrettable that those in the senior ranks of the company are often those "chosen" to attend the shows, particularly the international shows. They are the personnel most remote from the tactical aspects of the new food product development process. They rarely attend the technical sessions; they never report back to their staff or prepare a written report of developments noted; samples are rarely retrieved from the shows, and descriptive literature is "too heavy for the luggage".

The reader may correctly detect a note of frustration in my comments. The frustration comes from numerous experiences guiding senior management at food conferences and exhibits. Too often, these trips are looked on as rewards or perks, against which I have no argument. As long as the purpose of the trip, which is to be exposed to new food products, to gather information, and to see business opportunities for products, is kept clearly in mind, then why particular individuals are sent is no one's business. These trips should be hard work. They are expensive in terms of money and of key personnel's time. A company must expect, indeed demand, value for the money.

What preparation is required for those who attend? Prior preparation is required for all attending trade shows or technical conferences. First, time is valuable. Second, there is much to see and many people to meet. One must be selective. All personnel involved in the product development process should, whenever possible, preview the catalogues of products, the list of exhibitors, the abstracts of papers, or the list of attendees. A shopping list of information, contacts to make, samples of products to obtain, and papers to attend for those not able to travel can then be prepared beforehand. With this in hand, those chosen few who do have the opportunity to go have a priority list to guide their activities.

2. Public Libraries

Public libraries may seem an unlikely source of food information or of ideas for new food products. Most public libraries, however, have sections on food and cooking with an extensive collection of cookbooks filled with recipes. Recipes detailing local, national, and international cuisines may be sources of ideas for new food products or may serve as starting points for bench-top test products. Cookbooks describing various ethnic cuisines provide ideas for developing or adapting traditional ethnic foods to cater to the needs of these communities.

Librarians, professionally trained in not only finding books but getting information, can direct one to specific information from specialized sources.

3. Specialized Libraries and Library Services

Specialized libraries, for example business libraries, reference libraries, technical libraries, and patent libraries, provide information and statistics on consumer trends in eating, on changing lifestyles, on new developments in technology, on health concerns, and on a host of other psychographic and demographic data.

The patent literature can be an interesting source of data. Food companies that are close enough to large centers where there are patent repositories should not fail to take advantage of the opportunity to send qualified personnel to investigate what is patented and what patents are pending. Here a company may find information revealing directions in research and development that competitors are pursuing.

FIGURE 5. Akara, a West African dish made from cowpeas, suggested as a nutritious snack food. (From McWatters, K. H., Enwere, N. J., and Fletchers, S. M., *Food Technol.*, 46(2), 111, 1992. With permission.)

Technical libraries are equally profitable in providing a company with ideas for new products or new developments in the science and technologies of foods, food processing and preservation, and nutrition. For example, here is a sampling of new product formulation, processing, and test marketing, a fraction of what is available, that was gathered from the technical literature:

- Mazza (1979) describes the complete process for extraction of juices from saskatoon berries, chokecherries, and rose hips and their manufacture into jellies. Also described are marketing and consumer evaluation studies of these native fruit products.
- Canned carrot pie fillings are described by Saldana and co-workers (1980). Nutritional data are supplied and compared with data for pumpkin pie filling. Sensory test evaluations are included.
- A nutritious, chocolate-flavored shake-type beverage prepared from chick peas and its processing are described by Fernandez de Tonella et al. (1981).
- A nutritious snack food, akara, prepared from cowpeas, was developed and consumer tested by McWatters et al. (1990; 1992). This is a popular West African dish (see Figure 5), which has been well received as a snack item in affective testing trials.

- Prickly pear fruit was used by Ewaidah and Hassan (1992) to prepare a fruit sheet. Formulation, composition, and sensory evaluation are presented.
- A step-by-step procedure for manufacturing fruit bars (fruit leathers) is described by Amoriggi (1992). He includes information on formulating mango, banana, guava, and mixed fruit bars. Packaging and storage information is included.
- Saca and Lozano (1992) describe a process for explosion-puffing bananas to produce a dried product suitable for snacking or reconstituting in bakery goods.

Specific ideas such as the above may not have direct interest for a developer but may provide guides for formulations or processing or suggest products that can be adapted from these.

For the developer interested in chilled foods, the discussion of spoilage of chilled ready-to-use carrots, by Babic and co-workers (1992) may guide the design of stabilization systems that must be built into such products. Best (1992) presents an excellent working review of a new and somewhat controversial view of water in relation to food components in product formulations. The article (Slade and Levine, 1991) that originated this controversial view introduces an interesting concept of water in food that has implications for the quality and safety of many foods stabilized through the control of their water content.

Careful reading of any of these articles by product developers could be the source of many exciting new product ideas.

The value of reviewing scientific and technical literature is threefold:

- First, the subject matter of articles has value for what it describes about some food or process.
- Second, information on the authors and where the work was performed provides contacts with whom the product developer may wish, in the future, to communicate.
- Finally, there are acknowledgments at the end of the article of the supporters of that particular research. This information, plus the authors' addresses, may identify who sponsored the work and thus reveal competitive activity in a particular field.

There are computer-assisted information retrieval systems (see Chapter 5 for a more detailed description of these), which give subscribers access to various databases to fit any need for information. With a computer hook-up into the database, a subscriber types in key words describing topics of interest; the desired information is sent to the subscriber's screen. Information can be printed from the screen. These services are reasonably priced and can complement the resources of small library collections.

4. Trade Literature

Ingredient and equipment suppliers provide trade literature, newsletters, and bulletins describing their new products and their applications. This literature can be surprisingly fertile ground for new product ideas or information that will affect new product development. For example, Jones (1992) surveyed food labeling directives for EEC flavoring in a very timely review published in *Dragoco Reports*, which is invaluable for developers who may be contemplating entering the EEC with flavored foods.

Ingredient suppliers, in particular, often provide sample recipes employing their products. These are excellent sources of scaled-up formulations accompanied by processing details. The California Raisin Report, Winter 1993 issue, describes kosher foods and provides recipes for various baked goods suitable for Jewish traditional holidays. The American Spice Trade Association supports a column, "Flavor Secrets", published in *Prepared Foods*. The March 1993 issue, entitled "'Comfort' Pies" describes meat pies from various countries, provides a recipe for one, and advertises the availability of pilot recipes for others. Variations on the meat pies (miniaturization) could be the beginning of some interesting finger food snacks. Ingredient sales representatives are very helpful in providing specific information for new product ideas.

5. Government Publications

There is a wealth of new product ideas in the deluge of literature that is available from governments and their various departments and agencies. Governments frequently want to promote the use of agricultural commodities or underutilized crops. They provide recipes utilizing the foodstuffs, manufacturing directions, and occasionally market test data. No food company can afford to overlook this free and readily available source for increasing its ability to generate new food product ideas.

Government publications are also valuable sources of much demographic data, such as population movements, age composition of the population, incomes, food and nutrient consumption per day (Anon., 1980b), meal patterns, and so on.

Public information disclosure on who has received research monies or development loans provides information on activities in food research and plant construction among competitors.

Descriptions of developments in regional food research laboratories and agricultural research stations are published regularly. For example, the Food Research and Development Centre at St. Hyacinthe, Quebec, Agriculture Canada, publishes such an information bulletin. In one issue, Gélinas (1991) describes work on frozen bread doughs. The National Academy of Sciences (U.S.) (Anon., 1975) published a book describing the properties of underexploited food plants with promising economic value. Included in this are descriptions of the vegetable chaya, now readily available in supermarkets;

winged beans; the cereals quinua (quinoa) and grain amaranths, both of which have become more commonly used products; and the oilseeds jojoba and buffalo gourd. Specialty stores in many large cities have all of these presently available. Another example from the Food Development Division (1990), Agriculture Canada, is an extensive study on modified atmosphere packaging for the developer interested in using this technology for the development of new products.

Government literature should not be overlooked as a source for ideas.

C. Total Marketplace Analysis

The more developers know about events in the marketplace, the more accurately they can define their consumers and their consumers' needs and expectations. Generating ideas for products to fill those needs becomes much easier.

The marketplace must be scanned carefully for all the new product ideas that can be derived from it. The consumer is in that marketplace; the consumer influences that marketplace and is influenced by it; knowledge of the marketplace will help the developer to know and understand the forces shaping the consumer's needs and desires.

1. The Competition

It is a sad truth that competitive products on the store shelves are probably the source of many new product ideas for many companies. It does make one wonder where the competition found their ideas!

However, it is a marketing axiom that a wise company knows what its competitors are doing. Any activity by competitors in the marketplace must be noted. Their activity may require some retaliatory action. The retaliation necessary can be the impetus for an accelerated product development program requiring the generation of new ideas for products.

Unfortunately, many food companies are unaware of what is going on in the marketplaces they service. Nowhere is this more apparent than on the supermarket store shelves. When companies do not know what is happening to their own products in the marketplace, it usually follows that they are equally unaware of who their competition is or how their products look and are judged by consumers against the competition's products (Daniels, 1993).

Food companies need to analyze the products and activities of the competition. To state the blatantly obvious, the competition competes for the consumer's attention and dollars. If competitive products are more successful, then why? Are there ideas there for improvement of existing products or for demonstrating a perceived edge of difference that is discernible by the consumer over the competition? Food companies owe to themselves the frequent analytical inspections of competitive products. Such inspections, including sensory and compositional analyses, can provide data:

- to compare ingredients (and to approximate ingredient costs and the competition's cost margins);
- to assess quality characteristics that appeal to the competitor's consumers;
- to evaluate flavor preferences with small taste panels;
- to rate package and label appearances.

Maintaining a strong competitive program of product development requires such activity.

In addition, information so obtained can provide ideas for new products. Companies can indulge in some free association of ideas to generate new product concepts. An attribute of one competitor's product put together with features of another combined with something from a third product may lead to a new product. Such free thinking can be very exciting for the developer.

2. The Marketplace

Food retailing is changing in response to consumers' needs and changing lifestyles. Corner grocery stores lost out to supermarkets but recovered as so-called convenience stores. Supermarkets have developed into conglomerates where privately owned bakeries, green grocers, fish counters, butchers, and so on are housed in a market-like atmosphere under one roof. Other trends in buying have appeared: as a response to the economy-minded consumer, warehouse clubs have emerged; the mail order food retail outlet has appeared for the exotica in food items (note any issue of magazines such as *Chile Pepper* or *Vegetarian Times*); there is tele-shopping for the stay-at-home shopper, a convenience for the elderly or incapacitated shopper. These new marketplaces will not soon leave the retail scene.

The perishable section, once confined to the supermarket's back wall, has grown in size to include a wider selection of fresh fruits and vegetables, salad bars for carry-out foods, as well as trimmed and cut ready-to-eat fresh fruit and vegetables (celery and carrot sticks, for example). Many of the larger outlets have arranged this section to simulate a fresh air market designed to channel consumers through the displays. Bringing an atmosphere of freshness and naturalness to the fore, in-store bakeries waft the delicious aroma of fresh bread throughout the store to complete the picture.

Consumers no longer feel guilty about not slaving all day in the kitchen. Rather, they now take pride in greater participation in family activities, in community affairs, or in self-improvement courses. To provide the convenience they need, retailers now stock a wide selection of delicatessen items in the chilled foods section, thus competing with the fast food chains. In the frozen food sections, easily prepared entrées vie for consumers' attention. Products are displayed to convey ideas for tasty and nutritious meal combinations for hurried consumers.

The wise developer looking for product ideas could well afford to spend some time watching people shop in supermarkets, in convenience stores, in mom-and-pop stores, in take-away food stores with their wide variety of finger foods, and in open air farmers' markets. It is in these marketplaces that new food product ideas can originate. In too many companies, neither senior management nor the new product development team has ever questioned their sales forces. What new food products do they, who see and talk with consumers and retailers, want to put on the shelves beside the competition's products?

3. Analysis of Purchases

Much information can be had about consumer shopping habits through the universal product code printed on foodstuffs and recorded at the time of purchase. Data can be obtained, such as what items are purchased as the total dollar amount purchased rises incrementally. Or which stores have the highest average purchase per receipt? What products are purchased with what other products? Ideas for new product development (or for some piggy-back couponing) may come from frequent purchase combinations noted. If product A and product B are purchased together frequently, would combining the two in a single product or packaging them together as a single unit be a logical product idea?

Such purchasing data based on product codes is available from market research companies. Obviously some caution is needed in interpretation of data of this nature: products purchased together are not necessarily used together. Nevertheless, studies of consumer purchases do yield ideas for new products.

A similar caution applies to the interpretation of data describing case movements of goods between warehouses. Much value is put on such data as indicating consumer interest in the products being moved. High volumes might incautiously be interpreted as great consumer interest. They may also mean only that the product is being moved around. Case movement is an indirect measure of consumer interest.

By studying all categories of products available in the marketplace, one can note gaps — products or categories of products not available. GAP analysis is another technique for generating ideas for product development. Simply put, marketing people (usually) select a particular product category and then examine the marketplace for empty spaces in that category.

In a typical GAP exercise, a grid is drawn. Each row of the grid describes some product attribute, such as texture, flavor, color, particle size, function, or application. Columns might be labeled, such as used above, i.e., solid, liquid, or gas, or ready-to-serve, condensed, solid (as in frozen), or solid (as in dehydrated) for terms pertaining to soups. When the grid is filled in with data from the marketplace, ideas for new products may be revealed by the empty spaces of the grid.

For example, if hot pepper condiment sauces were the category selected, marketing staff would construct a grid (see Figure 6). As the grid is filled in, the following would be noted:

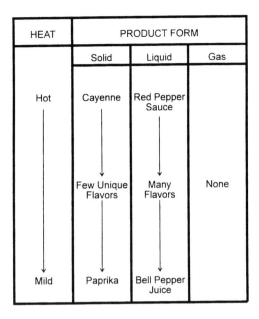

HEAT	PRODUCT FORM		
	Solid	Liquid	Gas
Hot	Cayenne	Red Pepper Sauce	
↓	↓	↓	
	Few Unique Flavors	Many Flavors	None
Mild	Paprika	Bell Pepper Juice	

FIGURE 6. GAP grid for hot pepper condiment.

- There were numerous liquid pour-on sauces, with wide variations in heat levels.
- There was a large number of solid "sauces", i.e., ground sprinkle-on products, with an equally wide variation both in heat levels and in flavor.
- There was no "gas" product, that is, pressure dispensed.

Here is a product gap. Does this suggest that there is an opportunity for a foam dispensed hot condiment sauce? (It could also mean there was absolutely no consumer interest in such a product, except as a personal protection device.)

In Figure 7, a grid for condiment sauces is presented. Horizontally, the form of the condiment sauce has been expanded from a liquid to a solid, with two possible forms: a concentrated paste (squeeze tube dispensed) and a ground powder. The gaseous forms are also divisible into two, dispensed as a foam or dispensed by gas as a liquid or paste, the latter pushing the product into more sophisticated packaging techniques.

Vertically, one can play with various flavors, as has been done here. Other sauces, based on other ingredients that have a hot principle, have been included. Some other alternatives would be placements, such as child-oriented flavors or adult flavors; time of usage, i.e., leisure time, as in barbecues or picnics; as a drink ingredient (in Bloody Marys or Bloody Caesars); or for formal meal occasions (canapes or gourmet sauces).

	Solid ⟶ Gas				
	Solid	Paste	Liquid	Foam Dispensed	Pressure Dispensed
Hot Pepper					
Ginger					
Garlic					
Mustard					
Horseradish					
Blends/Curries					
Black Pepper					

FIGURE 7. Expanded GAP grid for hot condiments in general.

GAP analysis, a form of attribute analysis, looks at the marketplace for product vacuums. Where a space is noted, no such product exists. No product may be there because consumers see no advantage for such a product or have no expectation for one. On the other hand, is there an undetected need for such a product, which has never been fulfilled because such a product has never been presented? The main purpose of this exercise, however, is to stimulate thinking on unfulfilled market opportunities as the grid is stretched beyond its limits both horizontally and vertically.

GAP techniques can be applied not only to an analysis of products but also to financial and marketing matters. Again a grid is made. Columns could be labeled as years when particular products (in the rows) reach the end of their effective life cycle. Or columns could be labeled as different sales regions of the country and rows could be the products that marketing needs to introduce into those markets to keep them strong. Or columns could represent anticipated income from different geographic areas of the country, projected for the next 5 years. Rows represent products both existing and new to fill the gaps opened up on the grid.

4.   The Consumer

Any and all information about consumers helps developers define the particular consumer to be targeted. As a result, the needs and expectations of that consumer can be better understood. Demographic and psychographic information about consumers is essential to this understanding.

Demographic information provides statistical data regarding population size and characteristics, such as the age distribution of the population, the income distribution, the number of family units as well as nonfamily units, the number of children per family unit, the male-to-female ratios, the ethnic background of people in geographic regions, the number of single parent families, and whether families rent or own their homes.

Per capita consumption figures are an example of demographic data. They are kept assiduously by governments. Commodity trade associations describe what is disappearing from the various market shelves and divide that amount by the population. They are not consumption figures per se, but are "disappearance" figures. They have only an indirect reflection on consumers' purchasing or usage habits and none at all on the habits of specific segments of the population.

However, while demographic information such as per capita consumption figures provides only bland information about consumers, other figures can offer real insight.

An astonishing amount of information can be obtained about consumers using both demographic data and various databases. This lack of privacy in the computerized age is a fact of life and a frightening concern for many people. General consumer information, as well as personal information, is sitting in databases all over the country. When put together, the information can provide surprisingly accurate pictures of consumers in general and individuals in particular. Where one lives and one's interests can be obtained from magazine subscription lists. An address provides information as to income status; access to property tax records identifies expensive areas. Membership lists of professional or trade associations identify occupation or financial status. In stores identifying purchases with scanners, data is amassed that can allow analyses of purchases, which when cross-referenced with data from other sources, can be very revealing about consumers, even individual consumers, and their habits.

One must remember that this demographic data may be very helpful but can be misleading if used in isolation. For example, according to demographic data:

- The over-55's are the fastest growing segment of the North American population.
- They have the most wealth of any age category.
- Females make up the greater part of that population since statistics show women outlive men. Consequently, many over-55's are either widowed or single.
- Women over 55 years of age need extra calcium to prevent osteoporosis. Other physiological changes in aging for all over-55's are general frailty, loss of taste buds, loss of teeth, and constipation.

Some product developer may incorrectly see an opportunity here to bring out a

- high calcium ... (older women have low calcium intakes and suffer calcium depletion),

- low sodium ... (high blood pressure is common in older people),
- low fat ... (the percent of calories in the diet due to fat is acknowledged to be too high),
- added-value ... (older people can pay), bland, soft food fortified with fiber and vitamins (for good measure) in an easy-open (over-55's lose their strength and dexterity), single serve (they live alone) container.

Can all over-55's be grouped into such an uncompromising classification?

Such narrow thinking could only lead to a new product failure. Certainly there are frail and incapacitated over-55's in nursing homes and other institutions. For them new products will come from astute product developers in food service companies. But the over-55's are also those finishing their schooling that was interrupted by children. They join travel groups and travel across China. They initiate their own Olympics for the elderly; they consult either professionally or voluntarily to small businesses or to projects in Third World countries.

The aging market, as one example, must be seen clearly to develop ideas for it. This market is highly segmented, and to capitalize on its vitality, the generation of new product ideas must come from knowing these consumers well in all their many facets. More than demographic data are required for this understanding.

Psychographic data reflecting behavior and attitudes of consumers must be obtained. This provides a better definition of the qualities that must be designed into products to meet the needs and expectations of targeted consumers. Psychographic data together with demographic data provide the inspiration for ideas for food products. More leisure time (demographic data) and a desire for a healthier lifestyle (psychographic data) may suggest ideas for nutritious snacks or low calorie beverages, such as fruit beers, wine coolers, or low alcohol beers, to accompany leisure activities. In the 1960s and 1970s consumers were concerned, in general, about nutrition. Nutrition was, in some vague way, good for one's well-being without consumers being very sure why. Now consumers know why. Today consumers focus on very specific health and nutrition issues such as:

- their diets and foods that may contribute to heart disease or may improve their circulatory health,
- their diets and foods whose consumption may contribute to cancer and, just as important to the consumer if not more so, foods that protect against cancer,
- diet and behavior.

At the same time, consumers have not abandoned such concerns as low calorie, low-salt (sodium), high-fiber (the good kind), low-cholesterol, and natural foods, by which consumers generally mean additive-free foods.

Consumers are slowly changing cooking and eating habits. Consumer magazines are enjoining their readers to fry less and broil more. A wider selection of many different varieties of fresh vegetables is available. These are becoming a more important and attractive part of meals, pushing, but not yet supplanting, meat as the main item. Fresh fruit is eaten more frequently.

Meal patterns have changed. In Victorian days or earlier, four and five meals were the order of the day. Our modern world reduced that to three. However, observations of food habits of most office workers today show that five and six meals are more common than the three "square" of more recent times. From discussions with students and with staff from a period when I was vice president of technical services for an international food company, the following meal periods with typical fare were common with both students and office workers, particularly young female office workers:

1. Morning: breakfast was fruit juice, coffee or tea, cold cereal or toast, or muffin. The beverages and the muffins were frequently eaten in class or at their work station.
2. Midmorning: they chose coffee or milk or juice, Danish pastry and/or muffin and/or fruit.
3. Noon: lunch was sandwiches, yogurt, hamburgers or hot dogs (the students' fare), cole slaw, french fries or poutine (french fries with curd cheese smothered in a spicy sauce, a choice that is, if nothing else, filling and warming in the wintertime for students).
4. Midafternoon: fare was similar to midmorning, with the possible addition of machine-dispensed foods such as soft drinks, peanuts, and chocolate bars.
5. Late afternoon: students relaxed over beer, chips, and other snacks; office workers went home to a light deli snack (paté, cole slaw, bread) before going out for the evening.
6. Evening: students ate supper in the cafeteria or at apartment residences and then went on to studies.
7. Late evening: both office workers and students ate some snack before retiring.

Perhaps the word "meals" should not be used and instead "eating periods" or "snacking periods" should be used. Today many people graze. They follow no traditional meal pattern regulated by the clock but eat when they are hungry. Finger food (much of which is ethnic in origin) and nutritious snacks are a natural consequence of this change in eating habits.

Development takes time. If, from demographic data, it is predicted that the population in a given market area will drop by a million people over a period of four to five years, then the consequences are that a million meals or meal items will be lost each day each year. That is an important fact of life for any product developer, food marketer, or food company. A company has only 4 years to find new products to fill the void caused by the loss of those meal items. Four or five years is not an unusual gestation period for the development of a highly innovative new product. Consumers who might be expected to buy that innovative product are several years older than they were when the developer originally conceived of the product.

Time is not on the developer's side. Developers are not aiming for today's consumer. They are developing for some consumer 2 years or more in the future. For these reasons, when products for development are conceived, they must not be based on today's demographic and psychographic data, which itself may have a time lag in it. The data, today's data, must be extrapolated to describe the consumer of the future.

Sometimes changes can be very rapid and unstable. Products based on short-lived fads must be identified as such early in the development process.

All of the preceding discussion has been devoted to one driving force — getting to know consumers in order to be able to generate ideas for new products that may satisfy those consumers' needs and expectations. Data, when interpreted, generates information. Information, in turn, generates ideas. Quantity of ideas, not quality, is important at this stage of the process. Critical screening will eliminate bad ideas later.

The generation of product ideas must come first, from within a consumer's mind, so to speak. This requires an intimate knowledge of consumers. The first net for ideas is cast wide. As both demographic and psychographic data about consumers are gathered and converted into information, the net is pulled tighter. A picture of the needs and expectations of a very specific group of consumers will emerge.

Data and information are available to either small companies or large companies. Management of many small companies believe large companies, with all their resources, have an advantage over them in idea generation. On the other hand, management of the large companies often look with envy at the flexibility of the small company, the rapidity with which it can respond to trends, its closeness to consumers, and its absence of bureaucracy and red tape.

Small companies can attempt the more unusual. They have less to lose. Failure of a product is less damaging to a brand image, less noticed by the consumer. Large companies, with more to lose by the failure of the launch of a new product, tend to be more conservative. Their managements are more influenced by peer pressure to conform and to "go by the book".

III. GENERATION OF NEW PRODUCT IDEAS: REALITY

This discussion has attempted to show that the means to develop ideas for new product development are well within the capability of all food companies with the desire to pursue them. The rub is, of course, do they?

In 1983, Goldman surveyed food companies (47 companies ranging in size from those with fewer than 100 employees to those with more than 500 employees) in southern Ontario for their management practices with respect to food product development. The companies processed foods by canning, freezing, bottling, drying, and chilling, as well as by packing fresh. The products processed were meats, fish, cereals, fruits and vegetables, beverages, dairy products, and confectionery and pet foods.

Goldman (1983) found that the method most used for idea generation was to imitate products of competitors in the marketplace. This leads to products of the "me too" variety, with little innovation or originality and no thought for the consumers' needs and expectations. Techniques employing focus group discussions (Marlow, 1987; Cohen, 1990) and brainstorming sessions were next in frequency of use. (Focus groups will be discussed in greater depth in a later chapter.) Surprisingly, lowest for frequency of use were those techniques most easily performed or most readily available: attribute testing, recipe books, company personnel suggestion box (asking the sales force what new products they would like to sell), and the patent literature.

In companies with a formalized organization for new product development, Goldman (1983) noted that as the level of organization increased, there was a greater tendency to make use, regularly, of a wider variety of techniques for idea generation. This might also suggest that where companies employed a more disciplined approach to new product development, idea generation was considered a more valuable tool.

The disparity between what idea generation could and should be and what it is in reality (Goldman, personal communication, 1993) is enormous and somewhat terrifying. Companies put money and effort into new food products, which are wasted in the failures seen in the marketplace. Following a competitor onto the store shelves with a "me too" product is risky business at best. The competitor has, it is hoped, researched the marketplace to obtain a clear picture of the consumer's needs and expectations. The manufacturer of the "me too" product does not have this picture. The competitor is into the market first, and if the product is a success and not just a fad, it will cost the copycat product more to get market introduction into the competitor's already established market. The originator has the processing know-how. The manufacturer of the "me too" product must learn the technology, and time is not on this company's side.

Looking at the competing products on the shelves and analyzing a competitor's product are methods that should be used by a company, not for copycat products but for ideas that will compete with the next generation of products aimed at the constantly changing consumer. No company can spend too much time generating ideas based on all the information that can be gleaned. Shrewd screening will weed out bad and unprofitable ideas later in the process of new product development.

Chapter

3

Screening for the Better Ideas

Ideas for new food products have now been gathered from every source imaginable. Some ideas may appear very logical on the basis of initial marketplace data. At first glance, others may appear to be quite bizarre. However, the wildest ideas have an element of brilliance if shaped by skilled personnel dedicated to the growth of the company. The next phase is sifting through these ideas to select those best suited to satisfy goals important to the company. The sole purpose of screening ideas is to improve the odds of success of a new product in the marketplace. Screening promotes the advancement of those product ideas most likely to meet the needs and expectations of consumers and to satisfy the goals of the company for growth with successful launches of new products.

Screening does not eliminate ideas. No idea should ever be thrown out. Ideas may be inappropriate for further development for any number of today's reasons. However, in the future, the reasons for having abandoned those ideas may be no longer valid. Problems encountered in the past become surmountable in the future. All rejected ideas should be recorded, with the reasons for nonpursuance stated, cross-referenced and then filed.

Several elements are encompassed in screening new product ideas. First, there must be an organization with a leader to coordinate the separate tasks of development. This leader has the responsibility of deciding whether to advance any given idea for further exploration. This individual is the product development manager: the body of the organization will be the manager's new product development team.

Second, the new product development team must have clearly stated company objectives. These guide it in evaluating progress through subsequent phases of development. The company's needs must be written down to establish clearly the goals to be reached within a specific time frame and within stated costs. Criteria can then be developed that the team can use to assess how closely products in development meet these objectives. The team cannot operate divorced from the financial, strategic, and tactical planning of the company.

Third, and finally, the new product development team requires laboratory, test kitchen, and pilot plant facilities or access to these. As well, there must be marketing skills to explore ideas for products with qualities that meet the needs and expectations of consumers. If a small food company has no such in-house capabilities, new product development companies and market research companies are available that will develop products and research markets for a fee.

Screening is not a one-time incident at the start of the development process. Development takes time. Six months, a year, five years, or more could elapse, depending on the nature of the product under development. During this period, the marketplace is changing. If nothing else, the originally targeted consumers are older. Screening proceeds throughout the entire new product development process.

In this chapter, this team, its objectives, and the criteria used to evaluate ideas will be discussed.

I. THE NEW PRODUCT DEVELOPMENT TEAM

The new product development team should embody skills necessary to evaluate ideas that justify further exploratory studies and to screen all subsequent work. Their resources should include the following:

- management skills
- engineering skills
- financial data
- production advice
- legal advice
- research and development
- marketing/sales data
- purchasing input
- warehousing/distribution data
- quality control advice

Food companies, depending on their size, have some or all of these skills and resources available to them within the departments housed in their organizational structure. Those skills that are wanting in-house can be readily obtained through private outside companies.

In small food companies, staff members wear several hats. Plant managers double as plant engineers; quality control managers frequently are responsible for research and development; even presidents serve other roles, perhaps as financial officers or purchasing agents. In the small food company, therefore, development teams are smaller, and they may need to contract outside skills and resources. For example, lawyers or accountants may be retained or market research firms may be hired.

Communication among a small company's team members is usually good. The greatest danger in such a closely knit setting, however, is that one strong-willed individual — the company president/owner, perhaps — will dominate. This can be a challenge to unbiased screening.

By contrast, larger companies usually have all these resources in-house. Several new product teams may be working independently of each other on a number of different projects. Each has a product manager, who can be considered the recording secretary of the team. This manager reports up the line of communication, which can involve several levels of more-and-more-senior management. The various teams rarely intercommunicate. As a result, communication can be more difficult among individuals of different teams, both horizontally and vertically. Frequently duplication of effort occurs.

As work proceeds, dominant roles within the team will vary as major activities in the development process vary. At some phase, food technologists dominate as they develop recipes and prototype products. At another stage, marketing may be conducting sensory evaluation sessions on small consumer panels while other team members await their results. However, none of the individuals with the skills that they possess ever stops having an input into the team.

Development teams usually remain in place until the project has been released to production as a part of the company's product line. This movement of the individual team members with the project gives the members a greater appreciation of all aspects of product development. They see the whole picture and understand the complexities of product development.

A. Management

Senior management is an integral and important part of the product development team, if for no other reason than to see and be seen. Management serves many diverse purposes:

1. It establishes clearly the interest of senior management in the project, which can play an important role in encouraging team spirit.
2. Management can establish that the company's objectives are being strictly adhered to and divergent paths are not dissipating the energies of the team.
3. It allows senior management to ensure that ideas selected for development fit the corporate (or brand) image of the company.
4. Management's presence can hold in check the rivalries, and sometimes abrasiveness, that can arise as pressures and deadlines take their toll of even the most integrated venture team.
5. Management has an opportunity to assess the strengths and weaknesses of individuals possessing different skills as they work under pressure with other members of the team. This allows a just reward for good work — in itself a morale booster — and permits management to earmark a cadre of future leaders.

The above does no more than tabulate the responsibilities of good managerial stewardship.

In small companies, presidents are normally intimately involved with their companies' growth and development plans at all stages. Goldman (1983), in her survey of the product development management habits of food companies in southern Ontario, found that in 45.4% of companies with 99 or fewer personnel, the president claimed the main responsibility for product development.

As companies increase in size, the president's involvement in product development becomes more remote. His place is taken by more junior staff. Only 12.5% of presidents of companies with more than 500 employees affirmed that they had the main responsibility for product development (Goldman, 1983). In large companies, the product manager serves this function and maintains a close liaison with senior management, usually at the vice presidential level.

In these larger companies, a more organized approach to product development was noted when 37.5% survey respondents indicated their job function as product development. No person from a small company who filled out the survey questionnaire described his job function solely as "product development" (Goldman, 1983).

The intensity of the involvement of senior management obviously varies widely with the size of the food company. Usually the greatest involvement is in the initial screening phases; this tapers off until the time when go/no-go decisions are to be made.

B. Finance Department

The financial member of the team oversees the costs of the project. Project costs need to be tracked carefully to alert all members as to whether the project is within budgetary limits. Marketing personnel, with this cost information, can compare the projected introduction costs with sales predictions. The result, it is hoped, is the net profits that the new product will earn.

The relationship between developmental progress and costs can be seen more readily in Figure 8. Here, in the upper graph, the number of ideas under consideration or in development is charted against time. The numbers decrease during preliminary screening and the later phases of recipe development. Sensory testing, consumer researching, and production scale-up weed the ideas out until few are left, and finally one is selected for introduction. The upper graph continues after a break, but now sales volume of the product being introduced is plotted against time.

The lower graph plots money flow (costs are –$: profit is +$) against the same time scale as the upper graph.

Costs (–$) are minimal in the early phases of screening (Figure 8). It costs little for development teams to evaluate ideas with readily available marketing,

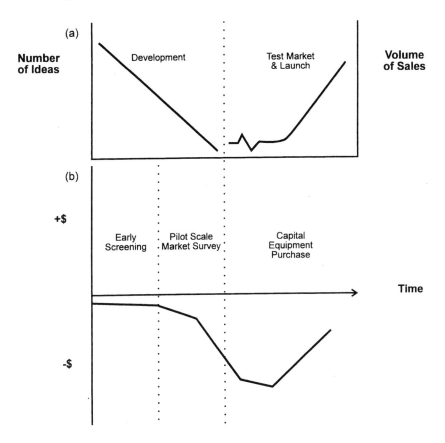

FIGURE 8. The relationship between (a) the course of development and (b) the costs of development (–$).

production, technical, and financial data. When preliminary work begins on screening formulations and if outside marketing and consumer research companies must be hired, expenses (–$) can mount rapidly. Then sensory testing begins on a larger scale. Many plant trials using expensive raw materials and disrupting routine plant production may be necessary.

Costs (–$) increase as development proceeds and researching techniques become more sophisticated. New equipment may have to be purchased. Costs for minitest markets and their attendant market research increase the amount of red ink. If new equipment must be designed and fabricated, a steep increase in costs (–$) can be expected.

If the test market is successful, a wider launch is justified. Eventually, sales volume overcomes the red ink. Figure 8, it should be noted, is in reality the "front end" of Figures 3a and 3b.

It is very important that project expenses are monitored closely to assure that they do not get out of control. Development costs can be a factor for dropping a project, as will be seen later.

C. Legal Department

The legal implications in product development cannot be overlooked by the product development team. Some member of the group must be knowledgeable about all the legalities that might pertain to the products or the processes under consideration for development or must know where and how to obtain pertinent information. This knowledge will prevent the team from wasting time and money pursuing activities that are not within the law.

Government, its laws, and regulations arising from enacted legislation influence the food industry in many ways (see Table 1 in Chapter 1). In product development, legal issues will be a major factor in two very important areas:

- technology used in formulation and processing,
- labeling and marketing and promotion programs.

Good advice early in the process concerning the implications of food legislation will prevent costly surprises later.

Technologists recognize that food regulations have an impact on formulations. For example, some additives are not permitted in certain foods, or there are limitations on the levels that can be used. Standards of identity are in place for many foods; these impose specific compositional requirements. Meal replacements must meet nutritional standards. Technologists must back up nutritional claims made for the product with ingredients that will provide the stated nutrition. Product formulations must be screened against all legal restrictions.

Food scientists and engineers must be aware of the legal implications of patent (or copyright) infringement as new equipment is designed or processes adapted during development. Conversely, they must also recognize when equipment or processes they develop are patentable. Thermal processes must be verified for safety before being filed with the proper governmental agencies.

Packaging to protect products must be carefully chosen by technologists. There may be legal implications. Governments, very sensitive to environmentalists, have legislated against the use of certain types of packaging. Some governments have decreed that all packaging be recyclable or have required a pick-up program for empty containers.

Marketing departments must know their legal obligations regarding labels and trademarks, the guidelines for promotions and advertising, and the legality of the claims and statements they wish to make. There are unwritten laws, too. Not only can advertising not be misleading, but it must also not be sexist or racist. Package sizes are closely regulated.

Names chosen for new food products cannot be misleading. Goldenfield (1977) presents an excellent example of how food regulations can impinge on the process of product development. Goldenfield's example is a refrigerated whipped-fruit flavored puree, which the marketing department insists contains only fruit pulp and fruit juice so that advantage can be taken of declaring all natural ingredients. Naturalness is key to the concept. It is to be marketed under the product name "Fruit Fluff".

Problems begin when the technical team attempts to keep within cost parameters. The "FRUIT FLUFF Orange Dessert Whip" (name and product) undergoes a metamorphosis from "FRUIT FLUFF (Natural) Orange Flavored Dessert Whip" to "FRUIT FLUFF (Natural) Orange Flavored Dessert Whip With Other Natural Flavor" to finally emerge as "FRUIT FLUFF Artificially Flavored Orange Dessert Whip" as technical and budgetary considerations play havoc on the name choice.

New products destined for export markets bring further legal problems. Food laws of other countries respecting product safety, labeling, ingredient listing, permissibility of additives, and advertising must be adhered to. Reformulation may be required when export markets are sought for domestic products.

D. Marketing and Sales Departments

The most important responsibility for marketing and sales departments is to monitor the marketplace for changes that might affect the course of development. Their vigilance in the marketplace alerts the team to new considerations in screening criteria, to new market opportunities, or to necessary changes in products under development.

Marketing and sales staffs develop the marketing and advertising strategies for products under development. With the cooperation of technologists, they prepare label statements and claims, as well as recipe and usage suggestions. They oversee art work for labels, copy for advertisements, and media campaigns that will be used in promotions.

Marketing personnel consider what impact new products may have on a company's established branded products. Will there be fragmentation of the market? If so, is this fragmentation good or bad? Within the established market there may be several smaller markets — niches — that might be exploited profitably. The instant coffee market was fragmented profitably with the development of flavored coffees, for example.

Marketing and, in particular, sales members of the team must evaluate what impact products may have on the retailer. The following example illustrates this concern. A microbrewery (craft brewer) introduced a prestige beer with an old fashioned, wired-on cork stopper. Minitests were conducted in nearby campus pubs. The container was redesigned after complaints from bartenders and waitresses poured in. Opening bottles was inconvenient; service was delayed; hand injuries resulted. Wires proved a hazard underfoot, and the corks made good missiles. The beer was fine.

Retailers and their concerns can be a significant factor for the marketing and sales group to consider in screening and development. If a product goes on the shelves, how will the retailer react? What other products may have to give up space?

E. Warehousing and Distribution Departments

Warehousing and distribution staff must have some input to the screening process. Their considerations could determine the direction of development because of the cost factors they introduce.

Special warehousing or distribution channels (chilled or frozen foods) or special environmentally controlled storage may be required for either ingredients or finished products. For example, popping corn and chipping potatoes require special temperature and humidity storage. Many fruits and vegetables require controlled atmosphere storage to keep adequate supplies on hand. If these are not available in-house or locally, then the cost of obtaining these additional facilities must be factored in as part of the development total.

Considerations must be weighed for the need to have a returns pick-up for short shelf life products kept beyond their expiration dates. There must be a pick-up service for recyclable or reusable empties. Warehousing and distribution personnel must evaluate, as an integral part of screening, both the resources and the skills needed to make adequate storage and distribution systems available.

F. Engineering Department

Engineering personnel play a key role in screening. They evaluate processes needed for the manufacture of new products. If new or innovative processes are key to the product's success, engineers may have to modify old equipment extensively or design novel equipment for any innovative processes involved. They can identify the delays, costs, and disruption that will be involved in development and in normal plant operations for such activities.

Late recognition of the increased burden new products will put on existing processing lines and physical plant can disrupt the timeliness of their introduction. This can be disastrous for the launch of seasonal new products.

Engineers should work closely with production personnel and food technologists to design safe processes for plant scale production of products. These will be needed for the many consumer studies that will be required before test marketing.

They research equipment suppliers for delivery dates and costs of needed equipment. New production equipment costs or delays that may be encountered in delivery of equipment must be communicated to the team early as a consideration in screening. Six to eight month delivery dates are not unusual for needed machinery.

G. Production Department

The production staff's role in the screening and the development process has already been referred to indirectly. As early as possible, they must communicate whether the necessary skills, labor, and physical plant are available to produce the new product within the cost constraints and quality parameters required by marketing.

They work closely with engineers to evaluate available processes and to determine what modifications may be necessary. Between them, they must communicate, as their contribution to screening, the impact that the new product ideas will have on the existing plant production systems. They signal the need to engage a copacker for the new product or to purchase equipment for self-manufacture. These decisions have a decided impact on product costs and the rate of return on investment.

Production staff work with marketing staff to review their sales volume projections and to coordinate manning requirements for the processing lines. Their knowledge of local labor markets and the availability of skilled labor become factors in screening deliberations if a high level of technical skills is required for the manufacture of the product.

H. Research and Development Department

Food technologists have a vital role in screening all the way through the development process. They control the technology of the development process. It is their skills that determine whether quality objectives espoused in the product statement can be met. Food technologists produce the bench-top models for initial evaluation by the team and by consumers in focus groups. From these, preliminary estimates of product costs are obtained. They further refine these bench-top prototypes until together with engineers and production staff, they plan plant-scale trials. With marketing, they plan and carry out consumer preference testing.

In test market, food technologists, as statisticians, may analyze the market data obtained. They will certainly evaluate the condition of the product during test marketing, both for the protection the package provides and for its quality of shelf life as determined during development.

They establish specifications for raw materials and the ingredients, in-line control standards, and the finished product, in cooperation with quality control personnel.

I. Purchasing Department

Purchasing departments have the unenviable task of finding inexpensive yet reliable sources of those raw materials, ingredients, and packaging materials specified by technologists. Where technologists have insisted on stringent or unusual specifications, purchasing departments may, in their search, have to submit many products that most closely fit the standards that technologists declare are needed. They must research several sources to ensure that continuity of supplies will not become a problem should national markets be opened if the product is a success.

The activities of the purchasing department play a role in product costs. The more cheaply the purchasing staff can obtain materials and ingredients meeting specifications, the lower product costs will be. The more exotic and stringent are the specifications, the higher the costs. With the traffic department, they will have to balance transportation costs with geographic availability, and the reliability of the source with the item's cost. This interplay between availability, cost, reliability of supply, and quality (adherence to specification of ingredients) of supply is one in which great cooperation and interaction is required between technologists and purchasers.

Product costs are a major screening tool. Too high a unit cost may cause rejection of a product because too high a price would have to be demanded from consumers. Purchasing personnel's success or failure in obtaining supplies is important to the success or failure of the project.

J. Quality Control Department

There has been much confusion over nomenclature regarding quality. By quality control, I refer to that function in a company responsible for assuring that all processing, product, environmental, and worker safety standards are adhered to and that all reasonable and practicable precautions to protect the product from hazards of public health significance have been taken.

Involvement of quality control personnel in screening and in the development process itself is important. Their demands, imposed by concerns for safety and wholesomeness, may prove too costly or too time-consuming to incorporate into the design of the product. The project may have to be abandoned or delayed while remedial development to overcome such problems is carried out.

Individuals responsible for this function serve as watchdogs throughout the development process. Potential hazards to the safety of products in the marketplace, or in the hands of consumers, as well as hazards to the quality of products, must be ferreted out and corrected or controlled. Quality control individuals screen by detailing those process requirements that will be necessary if products are to be safe and wholesome.

Now is the time when the development team has the opportunity to build into new products, while they are still in the development phase, all the elements embodying a sound quality control program to assure safe, whole-some, high quality products.

K. Summary

The foregoing descriptions of the roles of members of the development team show that overlapping of duties and responsibilities can occur. There are grey areas of responsibilities between the roles of engineers, food technolo-gists, quality control personnel, and indeed, production personnel. The over-laps can be, and sometimes are, a source of friction in the development process.

Management of these jurisdictional disputes is rarely a problem in small companies. In large companies there are sharp distinctions between depart-ments that can be jealously guarded. The potential for problems of communi-cation concerning responsibilities, either between different departments or individuals in these departments, can be real.

The point that is important to remember in management of the develop-ment process is that no opportunity should be missed to bring all the skills of the team to the process of screening. The team is not just a body of people representing departmental territories. The team personifies the skills needed for the continuous evaluation and screening performed throughout the develop-ment process.

II. THE OBJECTIVES

How new product development will assist companies to attain their objec-tives must be clearly stated before embarking upon a sound program of development. This demands that companies first have objectives. This, in turn, requires that companies know who they are, what business they are in, and where they want to go. When boardrooms are clear on these points, then a company's objectives can be stated clearly and be understood by all.

Companies are rarely so single- and bloody-minded as to want only to make more money. Greater returns on investment are certainly objectives to be desired by any company. But the driving forces are usually only partly finan-cial. Some other considerations may include:

- a desire or need for growth into geographically new markets to lessen dependence on the economic climate of local or regional markets, which may be dominated by the vagaries of a major industry,
- a desire to reduce dependence on commodity-type products and to in-crease profit margins and competitiveness with more added-value prod-ucts,
- a desire to expand a product base and reduce the dependency of the company on one or two products,

- a desire for greater local or regional market penetration with a broader range of products to maintain and protect a competitive position,
- a need to broaden the business base by ventures into other market opportunities (for example, a leisure food manufacturer venturing into deli type restaurant service).

This is by no means a comprehensive list of company objectives.

A company's objectives can be roughly classified into three broad categories:

1. those objectives that are purely financial, such as obtaining a greater return on investment,
2. strategic objectives, which can be both defensive and offensive in nature, to protect the company's market position against inroads by the competition or to gain market share from the competition and the result of strategic objectives,
3. tactical objectives, by which the company can chart its progress toward its goals.

These aims must be clearly stated and be known to the venture team members. Criteria they use in screening and selecting one product idea over another are always directed to the attainment of these objectives. If the plan of action is to go toe-to-toe with a competitor, either to gain market share or to defend an already established market being threatened by a competitor, then this objective can influence the type of product screened for and the criteria used in the screening process.

III. THE CRITERIA FOR SCREENING

Development is a complex system of screening steps that require some criteria (measuring sticks) to ascertain whether ideas should progress further. Does the project go forward or not? and on what basis?

Criteria for screening can be summed up by one word: capability. Does the company have the skills to do it? However, by probing deeper into what is meant by "capability", these criteria can be broken down further into:

- marketability, a descriptive reference both to the product and the company's ability to market the product, i.e., the company's skills. For example, highly innovative products require extensive education of consumers. Simple line extensions require heavy promotion against a battery of similar products on the shelves. Different marketing skills are required: the marketability of the two products is different;

- technical feasibility of the development process (can it be done?) and its corollary, ability of technologists to accomplish the technical complexity required of them in developing the new product;
- manufacturing capability of the plant;
- financial resources of the company to undertake the costs of the project with no guarantee of recovering its investment in an acceptable time, i.e. the risk it is willing or capable of taking.

Applying these criteria will be very difficult. Subjective, judgmental decisions are required. Unforeseen problems inevitably arise over how the criteria are to be applied and by whom. In large companies with their multiplicity of divisions and departments, leaders of new product development teams face challenges as friction may build between people or departments. Challenging people's skills can demoralize the team. Criteria must be applied evenly and justly to the screening process.

A. Marketability and Marketing Skills

Does the marketing department have the ability to market the product to the targeted consumer? Will the product require unique marketing skills or will these unique skills be required to reach the intended customer? A lack of adequate marketing resources within companies may make them unable to cope with new ventures. For example, not all companies possess marketing skills to introduce new products into the food service market or into food ingredient markets if they are not familiar with these markets.

Many markets can challenge or be beyond any company's marketing skills. Therefore, companies should be aware of these dangers in unknown areas. They should recognize their own shortcomings in their ability to market.

Can a market dominated by a single competitor be challenged? Is the marketing department able to penetrate a market with few major buyers or even a single buyer? How big is the market? Is it local? All these considerations should, with competent marketing departments, be uncovered in market research. Sometimes they are not. Yet these are very basic marketing considerations. They are useful screening criteria before development progresses too far. And, all these considerations have been implicated, with the benefit of hindsight, as marketing causes for new product failures (Anon., 1971). They are screening criteria which should have been applied.

Products that themselves lack marketability fall rather loosely into two categories: either they are so far ahead of their time that they cannot be marketed (in time they will be), or they are inferior to products already on the market, cost appreciably more than superior products already available, or exhibit a point of difference from existing products that consumers can neither perceive nor appreciate. These obviously should be screened out early in development.

B. Technical Feasibility

Technical feasibility of projects is measured by the project's chance of successful development, the time to reach successful development, and the costs of that success. The more qualitative attributes that are demanded in a product, the more research and development efforts are needed to attain these desired features. Greater effort in research and development, in turn, lengthens the time between the start of the development process and the final technical success of the project. Costs rise as a consequence.

Can development be achieved? How soon can it be done? What will it cost when it is done? These are questions important to team activities.

Marketing personnel need to have firm time commitments, so that promotional material, labels, and associated artwork for advertising are ready for the appropriate launch date. But before any launch date can be decided, the distribution channels need to be filled by distribution personnel. And prior to the distribution channels being filled, manufacturing must produce the product. But even before this, any special equipment must either be designed or be modified by engineering, or specifications must be written for new equipment to be purchased. Financial interests should not be overlooked here. Accountants need to have financial estimates for the project.

It is much like that children's nonsense song:

There's a hole in my bucket, dear Liza, dear Liza,
There's a hole in my bucket, dear Liza, there's a hole.
With what shall I mend it? dear Liza, dear Liza,
With what shall I mend it? dear Liza, with what? ...

Liza gives many suggestions to her companion, each one requiring a subsequent step, until finally the last step requires a pail of water, whereupon the entire rigmarole commences again.

Technical development of new products depends on the probability that technologists can succeed in matching claims implied in product statements with safe, stable products. Development teams enter now into the realm of probabilities and "guesstimates". Products with little or no creativity, products that are imitations of existing products in the marketplace, or products that are simple line extensions, can *usually* be brought on stream with a high probability of success in a comparatively short time. However, no development project is simple and without little hitches.

Each step in the development of products can be analyzed for its chance of success, its cost, and its time to success. Any one step can present an insurmountable hurdle and stop a project's chance of moving forward.

1. Probability (or Fun with Numbers)

Determining the probability of certain events happening is familiar to every student of statistics. For example, the probability of getting four heads when tossing seven pennies or the odds of picking a red ace from a deck of playing cards are common problems described in textbooks. Also familiar to students of statistics are problems associated with calculating the probability of a particular event occurring as the result of a sequence of events, when the probability of each step of the sequence is known. A typical example might be picking a black ace and then a red face card and then a black two. (Readers unfamiliar with probability statistics should see, for example, Bender et al., 1982 and Parsons, 1978, for a concise, readable account of probability statistics.)

If the chance of going from A to B in some sequence of events has a 9 out of 10 chance of success (0.9), there is a high probability that B will be reached. If there is a third stage, C, and the probability of going from B to C is also 0.9 then the probability of going from A to C, as any student of statistics knows, is

$$0.9 \times 0.9 = 0.8$$

Therefore, the likelihood that C will be reached from A is still high but somewhat diminished. If more steps are added, even though each step has a high likelihood of success, the chance for success becomes less and less from the original starting point A.

Instead of likelihood of success of an event or reaction, one could easily have substituted processing yields (Malpas, 1977). Thus, if in this simple processing sequence, a 90% yield were anticipated at each step, the yield of C would be 80%.

The a priori probabilities associated with tossing coins or picking playing cards from a deck of cards are either readily calculated or are determinable by a long series of trials. They can be established.

Problems arise when food product developers attempt to assign probabilities to phases of the development process to determine what the chances of success are. Objective probabilities determined from coin tosses and picking cards no longer apply. There is no history of observations from which one can state that, on the average, such and such an event will happen with a specific probability value. Rather, the developer is forced to assign probabilities that "... are arrived at by considering such objective evidence as is available and, in addition, incorporating the subjective feelings of the individual" (Parsons, 1978). Subjective probabilities assigned by developers to the various phases of development must be based realistically on the best available information. They must not be unrealistic probabilities based on an enthusiastic overassessment of the technological skills of the development team.

The development process for a hypothetical product has been broken down into a simple sequence (Figure 9). To proceed from a starting ingredient, A, to the final product, P, requires three intermediates, B, C, and D, and four intermediate processing steps. A, B, C, D, and P can represent any phase in the development process. They could be considered to be efforts to provide the

$$A \longrightarrow B \longrightarrow C \longrightarrow D \longrightarrow P$$

	A	B	C	D	P
Probability		0.5	0.8	0.3	0.9
Cost		$W	$X	$Y	$Z
Time		T₁	T₂	T₃	T₄

FIGURE 9. Probability, costs, and time factors during development stages for a hypothetical product.

desired flavor or the desired textural qualities or even the required shelf stability.

The intervening steps do not need to be physical alterations in the product at all. They could be the likelihood of getting a change in legislation for a permitted additive, or the necessary change in some product's standard of identity, or the possibility of penetrating a particular market. Whatever the phases are, they can be represented as logical steps on the way to products for which probabilities have been assigned.

Each step can be assessed a cost figure for its accomplishment. The sum of the costs, $(w + x + y + z)$, for each recognized phase in Figure 9 represents the total *developmental* costs to go from A to the final phase P. (It should be noted that these costs refer only to the costs of development or the costs of the processes involved. They do not recognize the impact of development on other areas in the company, such as sales or production, unless these have been factored in.)

The time to accomplish this sequence is estimated to be (T1 + T2 + T3 + T4), the sum of the subjectively assessed time requirements for each step.

The probability, the expected total cost, and the time expected for completion of the sequence in the project can then be compared with those for other projects. It must be remembered that they are all subjective estimates. These comparisons are considerations in the screening process.

What is the probability of success? The phases (Figure 9) range from a more than moderately difficult one (0.3), stage C to stage D, to the very easy last one from D to the final product, P, rated at 0.9. The overall success of the entire sequence, the probability of producing (or getting to) P, is a disappointing 0.1 arrived at in the following fashion:

$$0.5 \times 0.8 \times 0.3 \times 0.9 = 0.1$$

This poor probability of success, as well as the cost and estimated time to success, may suggest that abandonment of the project is the wisest move. Much depends on the company's objectives and its strategy to get to these goals.

If the product, P, is highly desired, the technology team may be tempted to tackle the difficult C to D process first. This may be the most economical approach to the problem that ingredient developers might choose, since it avoids the input of time and money in solving the initial phases if it should be determined that the project is not feasible within the time frame of the company at the C to D process (Holmes, 1968). Probability analysis does serve a useful purpose.

Again, if these were percentage yields in the manufacture of some new food ingredient, one would anticipate only a 10% yield for the entire process. If the product, for example some ingredient, is highly desirable, such a yield may be acceptable.

New and improved products are almost certainly to be successfully developed. For example, a breakfast cereal can be improved in several different ways. The probability of

> better flavor is 0.5,
> better crispiness is 0.7,
> higher fiber content is 0.8,
> longer shelf life is 0.6.

To get an improvement in this cereal with respect to one of the above quality characteristics — but without specifying which one — the chance of success is 1 (complete success) less the product of all the probabilities of failure or:

$$1 - (1 - 0.5)(1 - 0.7)(1 - 0.8)(1 - 0.6) =$$
$$1 - (0.5)(0.3)(0.2)(0.4) =$$
$$1 - (0.01) = 0.99$$

The product is almost certain to be a new and improved one.

2. Assessing the Projected Profitability

A number of rough rules-of-thumb or indices can now be generated to estimate the potential profitability of projects. They are, at best, crude tools based on good (or the best available) guesses. As arbitrary estimates of the economic viability of a project, they can be useful to counterbalance the intuition and "gut feel" that are frequently behind the unwarranted, continued support of many questionable projects.

The simplest and crudest technique is to compare total projected costs (development costs plus capital expenditures plus marketing and promotional costs) against the projected gross sales for the period within which the company wants its payback.

The very crudity of this technique underscores its shortcomings:

- Introductory promotional costs are highest when attempting to get market penetration for new products.
- Retailers show resistance to new products because these take up more of the already limited shelf space. Retailers want something in return.

- New products are notoriously less profitable per foot of facings, so that retailers are reluctant to stock them.
- The company may have unrealistic expectations of a satisfactory payback period.
- No new product is ever introduced to the market with the covertly implied intention of withdrawing the product in any foreseeable time frame. Therefore, the technique is unreal.
- Confounding all the above will be the competition's retaliatory action in the marketplace.

This measure focuses only on direct costs associated with new products. The impact that these products have in other areas is ignored. For example, the production department will have additional costs with added downtime for line changeovers and extra labor costs. Sales calls, whether by the company's own sales force or an agent's, will be more difficult with more product to sell in the allotted few minutes with the supermarket chain's purchasing agent. More sales people may be required to handle expanded lines. These all represent extra costs associated indirectly with the new product. Comparison of total projected development costs and total projected sales alone is not a reliable index of profitability.

Statisticians have come up with refinements that attempt to improve the predictability of success. One such, the profitability index (Holmes, 1968), compares the expected return to the total cost, i.e., return/total cost and multiplies this by the probability of success:

$$(\text{return/total cost}) \times (\text{probability of success})$$

This ignores all indirect costs.

The shortcomings of all arbitrary indices to predict success or profitability or market penetration are due to the imprecise nature upon which the indices are based. Predictions of sales by marketing for new products that have no proven sales record are imprecise. The indices make no allowance for the retaliatory action of competitors in the marketplace with respect to advertising, promotions, or price wars. Estimates for the probability of success, time for completion, and costs are exactly that, estimates. They are only as good as the information that went into their estimating. "Garbage in, garbage out", as the saying goes. The indices remain tools to assist decision-making and not tools to replace decision-making.

Malpas (1977) discussed the use of Boston experience or learning curves for what could be another criterion for determining whether research and development dollars should be spent. In general, when volume units of a product double, costs usually fall by approximately 20 to 25%. If this generalization does not hold, then it is time to seek a new process. Argote and Epple

(1990) discuss the value of learning curves in nonfood manufacturing and cite their value in pricing and marketing and predicting competitors' costs.

C. Manufacturing Capability

Manufacturing capability as a criterion in screening must be used judiciously. It is the least consequential of the criteria. Inability to manufacture new products may not necessarily be sufficient reason to cancel ideas for products. Manufacturing and packing capability can always be bought through copackers.

There is, of course, a price to pay. Copackers levy a fee per case packed. If new products are price-sensitive, companies will not have great flexibility in pricing structures because of the copacker's added cost per case. Lower profits result.

When employing copackers, a frequently overlooked cost is the extra vigilance for quality control. Companies will require quality control staff people resident in the copackers' plants to ensure that new products are manufactured according to specification. A technical person in residence at the copacker's plant during production is necessary, even when the most amicable and trusting relationships exist between the two parties.

However, copacking may be a small price to pay to launch products. If products should prove unsuccessful, no capital expenses have been laid out for equipment. If products are successful by meeting or exceeding sales forecasts, then plans can be made for plant expansion to undertake self-manufacture. Acquisition studies can be commenced to purchase the necessary manufacturing capability.

If copackers skilled in manufacturing the particular type of product can be found, development of new products can proceed. Companies know their specific costs. More to the point, development costs and time might be telescoped. For many reasons, use of a copacker can be a very attractive route to new product development.

Manufacturing capabilities and manufacturing skills should never be confused. The one, lack of manufacturing capability, is not an effective reason to stop new product development. A lack of manufacturing know-how, even when the physical plant is available, may be a very strong reason to abort a project.

D. Financial Criteria

Financial criteria for screening depend very heavily on what the company's objectives are. These vary widely with the stage of development of a company. For example, companies pursuing programs of new product development to overcome their dependence on seasonal processing have different financial criteria than companies developing new products to challenge a competitor.

For companies seeking an added-value product for an all-year operation, financial success is measured by less reliance on seasonal manufacturing, by developing a year-round operation, and by gaining a foothold in the marketplace with added-value products. They can be more patient in their expectations of return. Companies fighting action by competitors in the marketplace may measure financial success merely by having protected their market share or by assessing how much market share they were able to take from competitors. Their time frame is shorter.

As development progresses, the direct costs of development and the indirect costs associated with the new product's impact on the company's existing infrastructure become more apparent. More reliable forecasting of sales volumes based on consumer research should give clearer information of when recovery of expenses may occur (see Section III.2).

Financial criteria must be applied fairly. Company controllers can, by using accepted accounting techniques, assess certain expenses as assets (investments) and can regulate the rate of depreciation. All of this can reflect badly or well on product development, depending on the company's long-term view. Companies that favor short-term financial gain by cutting their development budgets risk losing managerial skills that could be their salvation in the future. Projects can prove a training ground for younger staff and allow more senior management to evaluate personnel for those who will be the company's leaders tomorrow. At the same time, in-depth investigations of specific projects provide valuable technical information to companies for reassessing future objectives in more esoteric fields of development.

E. Summary

Any of the criteria used to screen, i.e., marketability, technical feasibility, manufacturing capability, and financial criteria, could be reason enough to abort development at any stage. Two, marketability and technical feasibility, must be applied carefully and dispassionately. Probabilities of success in both these areas are based on subjective assessments. Those responsible for them can and do become emotionally attached to pet projects and are reluctant to accept the need to drop the project.

Being objective can be difficult for the product development team. Leadership must be enlightened and compassionate. It is, after all, in the "people business". At the same time, the leadership must be dispassionate in applying criteria in screening. It is here that leadership must be demonstrated.

From Product Concepts to Conceived Products: The Development Process

"A bird in the hand is worth two in the bush." Cervantes

Development is a progression from the intangible — the product idea embodied in a product statement or concept — to the tangible — the actual product with all the attributes stated in the concept, ready to be tested in the marketplace (the final screening). Food technologists begin by designing a product based on a concept statement and then alter the product according to the results of sensory and consumer evaluations.

Figure 10 is an attempt to provide an overview of the process (cf. Figure 4). The upper flow in Figure 10 depicts efforts that are largely the responsibility of the marketing department.

Something tangible, a benchtop or prototype product based on the product statement (the middle flow in Figure 10), is formulated by food technologists and used by marketing researchers to stimulate conversation in focus groups for further refinements of the concept or is used in minitest markets for the same purpose. With a prototype product in hand, sensory evaluations can begin to guide food technologists in refinements in formulation. Preliminary product safety and shelf stability testing may require further alterations both to the product and to the concept in these early phases.

The bottom flow in Figure 10 is largely the domain of the engineering and the production departments. Engineering and production personnel design processes for new products, incorporating changes determined by food technologists. At this stage, reliable sources for ingredients and raw materials should be found and product cost data confirmed. Potential copackers can be evaluated if it is thought that the product cannot be made in-house. As development progresses from left to right in Figure 10, there is, at any point, a constant interplay between the other two flows. Each shapes and molds the other streams.

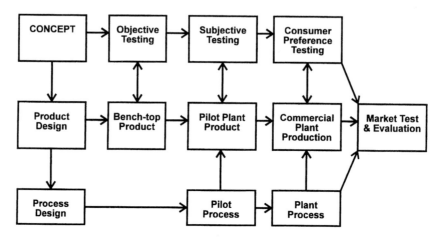

FIGURE 10. Idealized representation of activities flow in product development.

All of development, from idea generation, to test market, and even to a national launch, must be viewed as one long, on-going screening process. Each stage of development brings further data which, when translated into information, provides development teams with more refined tools with which to screen.

Rather inappropriately, screening is depicted in many papers and texts describing product development as a series of sequential stages in development, usually the second stage after idea generation. Holmes (1968, 1977) divides the development process into six sequential stages, Mattson (1970) chooses 11 steps to "low cost product development", and Meyer (1984) also gives 11 stages to successful new product development. Oickle (1990) describes seven stages, each of which has substages, while Graf and Saguy (1991) list five phases, each of which is subdivided.

Screening is not a stage in development; it is synonymous with development. As information is gathered from marketing personnel, from technologists, or from suppliers, the product becomes refined and sophisticated to meet the demands of consumers.

Blanchfield (1988) accurately describes the process as the continuing interplay of market research, technology, and financial efforts of companies to produce the right products at the right prices. With a series of very elaborate flow diagrams, he describes the process of development eloquently as

> a series of metaphorical "sieves" by a project team including representatives of the company's commercial brain (marketing), technical brain (product development technologists), financial brain (finance and cost accounting), muscle power (production) and nervous system (food control...), interacting via market research with panels of consumers representative of that market sector for which the product is envisaged.

Market and business analyses continue throughout all stages of development, seeking the "unexpected" with respect to volatile consumers. Food technologists, engineers, and production staff continue to refine processes. Information-gathering from the marketplace about consumers, about retailers, and particularly about the competition goes on continuously. Reliable information from anywhere may suggest to the product development team that the direction of development should be altered. Perhaps, for instance, a different consumer segment should be targeted.

The whole purpose of the development process is to move a product to market with the least amount of uncertainty respecting its probability of success in the segment of the marketplace where it is to compete. Market, financial, and technical research (and their associated analyses) are nothing more nor less than screening procedures for reducing, as much as possible, uncertainty in the development process.

I. DEVELOPMENT SEEN FROM THE MARKET SECTOR

All successful product development starts with the consumer. Many companies make a fatal mistake by starting with preconceived ideas of what they *think* consumers will want. They may very well be correct, but they also have a very good chance of being wrong. The misconception of what consumers want frequently has its genesis, especially in small and medium-sized companies (and occasionally in large companies), in a feeling, "a gut feel", which a member of senior management has for some product idea. "It will sell. I know it. I have this gut feel." In technology-dominated companies, the misconception may be that technologically advanced products will sell. Wrong! Consumers determine what will sell, if and when consumers' needs are satisfied.

A. Market Research

Market research is *an organized and unbiased investigation to measure qualitatively and quantitatively all factors influencing the marketplace and to provide information from the data gathered to guide marketers in decision making*. The subtle oxymoron "... organized and unbiased ..." stands out like a sore thumb. If an investigation is organized, some bias is introduced a priori by the researcher. It is the researcher who chooses which factors are selected for measurement and which are omitted. Decisions are also made on how to analyze and interpret data so obtained. These decisions necessarily introduce bias. Market research information will always have an element of bias in it. This does not damn market research. But only with careful and close liaison between the company's marketing personnel, the other members of the development team, and the consultant market research company (should one be selected to undertake this aspect of the process) can an objective assessment of the reality of the marketplace be obtained.

Market research starts early and continues throughout the development process. It *must* continue throughout development. The development process takes time. Consumers can be very volatile during this passage of time. If nothing else, they will certainly change by getting older. What were once thought to be good ideas for products today may prove to be short-lived fads by tomorrow when they are introduced into the marketplace. Such startling information is best learned early in the development process, before too much time and money are wasted.

The marketplace is replete with change, not all of which can be extrapolated with certainty to define trends. Experts abound, opining on the results of various surveys. Their interpretations of the results often diverge. There are very few facts to be found in the marketplace; observations, yes; facts, no.

Market research information is either qualitative in nature or quantitative. Qualitative market information is based on the interpretation of focus groups and interviews. Quantitative market information is based on the statistical analysis of data obtained from surveys and questionnaires. Quantitative research techniques aided by computer-assisted analyses are becoming very sophisticated. Techniques employed in researching the market (and there are many) require that the researcher clearly understand what information is wanted.

1.   Focus Groups

Focus groups, mentioned very briefly in the opening chapter, are qualitative information-gathering sessions. A focus group is an assemblage of consumers seated around a table, sometimes but not always with representatives of the client food company interested in the investigation present. Leading the group is a moderator, a trained discussion leader-cum-psychologist, who is employed by a market research company. Between 8 to 12 consumers, selected as representative of the target consumer that the company wishes to reach, participate. Marlow (1987) discusses the value of focus groups in product development, how best to use them, and the mechanics of using them. A caution is in order; they are best conducted by professionally trained moderators or farmed out to reputable consumer research companies.

Market research companies keep lists of consumers whose backgrounds are well documented. With little effort, they can enlist consumers with any desired profile that the client could wish for to participate in the focus group. The impact that a product concept might have on several different consumer profiles can be investigated by clients through these groups.

In practice, moderators will lead the discussion through some general topics and then, aided by some prop, stimulate a more directed (focused) discussion by asking leading questions concerning this prop. This prop may be a simple description of the client's proposed product, descriptive artwork depicting the proposed product, or actually a prototype product for the group to see. Moderators elicit comments from the group about this proposed product. They probe the responses and reactions of the group, always gently pushing to obtain more focused attitudes to the prop. Sessions usually last 2 to 3 hours.

There are as many variants of the above as there are practitioners of focus groups or market research companies. Proceedings may be taped and filmed to enable consumer research companies to analyze carefully the oral and body language used by the group for hidden clues in the responses. Several focus groups with different consumers may be required before a clear and concise concept statement for a new product can emerge that embodies what the product is and how it will meet the needs and expectations of consumers. Some consumer research companies use the same consumers over a 2-day session of discussions to get the desired concept statement.

All market research companies use focus groups as described above. The technique may masquerade under various, sometimes quite obscure, esoteric names. They all have in common a trained discussion leader who focuses the discussion of a group of consumers with a common background onto some food concept embodied in some prop. The leader later collates the reactions of the group to the prop and reports the findings to the client.

Done well, focus groups can provide valuable information concerning consumers' needs and consumers' interaction with the product concept or the mock-up presentation of the product. Cohen (1990) has discussed the advantages of using focus groups and reviewed their value, as well as the value of in-depth interviews in product development. However, interpretation of the data obtained is still very much a subjective analysis. Done poorly, focus groups can be a waste of money, for they are expensive. As Marlow (1987) remarks, focus groups have value for suggesting direction for ideas, but they cannot be the basis for business decisions. The main function of the focus group is to determine consumer reaction to products, and from this reaction, companies can redesign products to be on target with this reaction. A food company interested in using the focus group technique for market research should seek out a professional market research company with an established reputation.

2. Interviews and Surveys

Interviews are simply one-on-one situations between an interviewer and an interviewee in which attitudes of the interviewee are sought. A survey is a collation of a lot of interviews. There are two types of interviews:

- structured interviews, in which the interviewer has a very specific list to follow faithfully in posing questions to the interviewee,
- unstructured interviews, in which the interviewer will use a loosely structured prompt sheet with which to conduct the interview and record answers and opinions.

Cohen (1990) discusses interview techniques as a tool in product development.

Surveying the general public or specific populations within the public produces useful quantitative data for product developers. Whether a formal, structured questionnaire or a cue sheet of topics is used, this technique does permit, because of its personal nature, interviewers to get right to the heart of the matter. Unstructured interviews give great freedom to interviewers to probe very deeply into a variety of subjects; this permits the critical analyst to have greater insight into the mind of the consumer.

Interview techniques do have two very real weaknesses for market research — the interviewer and the structure of the questionnaire. Interviewers, using either structured questionnaires or prompt sheets, can in many subtle ways, consciously or unconsciously, sway respondents. Voice, clothing, body language, physical appearance, age, or sex of the interviewer can trigger this influencing effect.

Interviewers should avoid introducing any bias during an interview. This requires skill. Unfortunately, many interviewers hired to use the questionnaire for the less probing interview are college students earning money on school breaks. One-on-one interviewing is an expensive technique in market research: the use of trained interviewers adds to the expense. If the survey is improperly conducted, the cost is twofold; the money is spent on a poor interview, and inaccurate information is obtained.

The structure of questionnaires or their counterpart, prompt sheets, must be carefully designed to remove any ambiguity in the questions. The wording of surveys, the language of their preambles, and even the order of questions must be designed to avoid any bias that may be introduced in any type of survey, whether conducted by mail, by telephone, or personally. Consciously or subconsciously, respondents express opinions or answers directed or influenced by the wording in the questionnaire. Market researchers have an obligation to prepare and refine the questions put on surveys to obtain unbiased and objective data and not data that support preconceived concepts held by either the client or the market researcher. The wording of questionnaires to obtain objective responses is an art.

Data based on interviews and focus groups must be cautiously interpreted. Interpretations based on data from small samples of consumers should be extrapolated to general populations with reservations. The results, without hyperbole, should be seen as providing only a very general reflection of the targeted population.

Telephone surveys will reach a large population in a central market. They can also be used to reach a very dispersed population if the costs of long distance can be justified. In telephone surveys, questionnaires must be brief. Respondents are being imposed upon: interviewers have interrupted respondents at home. Questions, nonetheless, can be somewhat more detailed (given the constraints mentioned above) since interviewers can explain difficult questions. When computers are handy, interviewers can key in respondents' answers, and the results of the survey are available rapidly for analysis.

Again, bias can be introduced by interviewers in telephone surveys. However, it is much less in this situation, where facelessness removes some of the elements contributing to the bias.

Mail surveys have no geographic bounds. Since most marketing research companies keep mailing lists of respondents characterized by income, religion, ethnic background, and so on, great selectivity of respondents is possible for the mailings.

Mail surveys are the least costly technique of gathering data. They have no interviewer bias, but that does not mean that the survey questionnaire itself will be without bias. The explanatory text accompanying surveys, wording of questions themselves, or objectives of the survey can influence responses.

Printed questions must be clear and self-explanatory on mail surveys. Respondents have no contact for an explanation of ambiguous or complex questions. Any difficulty in understanding questions could either discourage respondents from finishing the survey or elicit false data. Respondents should be unable to misinterpret questions, so that wrong answers or impressions will not be given by respondents. This need for simplicity in mail surveys may limit the kinds of information companies can obtain.

An obvious disadvantage to a mail survey is that a less than 100% response rate is common. Response rates of 45 to 50% are considered excellent: lower rates of return are the norm. Frequently too, the slow return of completed questionnaires delays data analysis.

Mail surveys and telephone surveys both reach a larger and much more widely dispersed population of consumers than can be approched with either personal interviews or focus groups. One must remember that the type of information being gathered is, however, different. In Table 2, the advantages and disadvantages of surveys are tabulated for comparison.

Questionnaires mailed to specific addressees whose opinions are valued (for example, senior company executives) constitute an interesting type of mail survey used to great advantage in forecasting trends. It is called the Delphi method of forecasting. Questionnaires are sent out to selected addressees, soliciting their views on specific topics for which a forecast is sought. Selected respondents are acknowledged experts in the topics under consideration or are leaders in the various industries represented by these topics. It is much like conducting a think-tank by mail at this initial phase.

Opinions from the first survey are collated. A second questionnaire is formulated based on a consensus of the responses given in the first one. The same respondents are canvassed again but are supplied with the "expert" opinion derived from the first survey. They may or may not revise their opinions in the face of this new information.

Again, the respondents' answers are collated. The researchers can decide whether or not to continue with yet another questionnaire digested and distilled from the additional, "rethought" information gained in the second survey. Out the survey may go for a third time, and again, the answers, when returned, are

TABLE 2. Advantages and Disadvantages of Various Survey Techniques

Survey Characteristic	Type of Survey		
	Personal	Mail	Telephone
Cost[a]	High	Low	Low to medium
Geographic scope	Limited in area	Limitless	Limitless (long distance costs)
Potential for bias	High	Low	Low to medium
Availability of results	Slow to medium	Slow	Rapid
Selectivity[b] of respondents	Nonselective to highly selective	Selective	Low
Style of questionnaire	Complex, lengthy	Simple, medium	Short (is intrusive)

[a] Costs for any type of survey can be variable. Generally, personal interviews are more expensive techniques, especially if specific interviewees are targeted. Long distance charges can add to telephone surveys unless done locally in several cities.

[b] If respondents in each survey technique are preselected from screened lists, then of course, the survey can be highly focused. Random selection of telephone numbers provides no selectivity. Accosting shoppers in a mall also provides little selectivity.

collated. By this third round, a semiquantitative analysis — based on opinion, albeit so-called expert opinion — of future trends in the topics covered in the survey has been obtained. Food topics that might be considered for researching by the Delphi procedure include the following:

Where is the soft drink market going in the next 10 years?
What changes will occur in the fast food industry?
What impact will the green movement have on consumers' eating habits?

The Delphi technique of surveying is a very powerful tool in evaluating and forecasting trends. A good example of a Delphi survey is the Food Update survey described by Katzenstein (1975).

II. TECHNICAL DEVELOPMENT: THE CHALLENGES

Technical development begins when food technologists have a general idea of what the product will be. This may be before a clear and concise statement of the product and its attributes has been formulated. Only rough outlines of products-to-be are required before technologists can provide basic recipes for them. These are very easily obtained from cookbooks or ingredient suppliers or from an analytical breakdown of competitive products in the marketplace. These very basic recipes can start the food technologists on the path of determining the sorts of problems that might be encountered during development. The results from these early trials may be the bases for props used by marketing personnel to assist in refining concept statements with focus groups.

A. Recipe Development and Scale-up

These first cookbook recipes require much revision for a commercial recipe or for use in any larger scale pilot plant preparation. Owners of small entrepreneurial food companies frequently do not understand this. The president's wife's spaghetti sauce recipe or Mama's chicken soup recipe are not readily adaptable to commercial production techniques without extensive modification. Moreover, this lack of understanding of problems associated with scale-up of cookbook recipes exists not only with small family-run businesses. I have encountered it twice. Once, a president of a medium-sized food processor (his wife, actually) insisted that a particular canape spread could be processed as is, that development only involved increasing the quantities. Another time, this lack of understanding occurred with a prominent chef hired by a large food company. The chef was asked to formulate several frozen main course items suitable for factory-scale production. Both companies, incidentally, marketed nationally. In both instances, the items were impossible to prepare in quantity on commercial equipment with the available manpower and with readily available commercial ingredients within the cost parameters laid down for the products.

The need to modify family or cookbook recipes arises for several reasons, which require explanation for all nontechnical members of the development team as well as for novice developers. It is not a failing of food technologists who are unable to make products "just like Mama's". First, commercial products must be safe with respect to all hazards of public health significance and stable with respect to their high quality attributes throughout the entire distribution chain, right to consumers' tables. Certainly, products must be stable throughout their anticipated shelf life. Mama's product merely must be cooked and eaten within minutes of preparation; there is no packaging, no retorting, no handling by unloving, and more importantly, unskilled hands. And, no doubt, everyone in Mama's family is mutually immune from one another's infections.

Second, commercial plant operations must keep within well-defined product cost limitations. High cost ingredients used in many home recipes would prove too expensive for commercial practice. Cheaper substitutes adaptable to mass production must be used. Home prepared recipes have very flexible budget restrictions. Indeed, nothing is too good for family or for guests.

Furthermore, commercial plants have the extra expenses of labels, packaging, transportation, labor, advertising, and promotion, as well as the costs of support systems, such as quality control, plant maintenance, and plant sanitation to factor into costs; the homemaker's product had to include none of these. In the home, labor is free, experienced, and skilled. The only packaging required is a plastic container to store leftovers in the refrigerator. Of course, manufacturers must modify recipes to reduce problems (and costs) that could result from using the ingredients found largely in the home recipe.

Third, food preparation equipment in the home could never make sufficient product to fill the large volume requirements of commercial sales. The volume demands of the trade need to be filled by high speed food processing machines. To use these, the product in process must conform to certain specifications with respect to density, viscosity, particle size, uniformity, thermal properties, and many other variables required for reasons of safety, product uniformity, quality, and the ability to be handled well on high speed lines. The variability of home products would not be acceptable in the marketplace.

Nevertheless, some manufacturers are beginning to realize that consumers believe products with too uniform, every-unit-looks-alike-tastes-alike quality are synonomous with mass production and highly processed foods. Lightbody (1990) makes the point

> there is evidence to indicate that uniformity of appearance, texture and taste within some manufactured food products can be judged by some consumers as unattractive and, in some cases, a sign of 'excessive' processing.

One company with which I worked firmly believed that the wide popularity of their products was their variable, but always high quality, which gave their products the appearance of being homemade. They did not want uniformity in their pour-on sauces, and they cultivated nonuniformity by allowing a degree of syneresis and lack of gloss in their product. These imperfections made perfection in their opinion.

Finally, commercial products must meet the needs and expectations of consumers — not just Mama's family, or the president's wife, or a chef's sensibilities of what constitutes quality. No matter how good a cherished recipe is, to commercialize it and have it meet the needs and expectations of targeted consumers *at a price the consumer is willing to pay* requires much research and development on the part of food technologists.

The scale-up of recipes obtained from cookbooks to industrial kitchen batches to pilot plant batches to production-sized batches presents unique challenges to food technologists and food engineers. Products prepared on large commercial scale from recipes unmodified in any manner to suit the equipment they were prepared on rarely taste like products prepared in pilot plants and not at all like the kitchen-prepared products. Taylor (1969) suggests these problems are the result of any or all of the following causes:

1. Much of food technology lacks a sound scientific basis or, as Taylor (1969) puts it, technology is running ahead of the science. When this occurs, "scale-up is not very soundly based."
2. Raw materials used in food processing are highly variable from one variety to another variety of raw produce, from supplier to supplier,

from season to season, and during the season, from one geographical area to another. One thing engineers dislike in designing scaled-up plant is variabililty. Overcoming variability in processes requires operator skills.

3. Pressure mounts to cut corners in new product development to meet launch date targets. "There is regularly pressure by marketing to eliminate any intermediate scale-up on the grounds of time saving. Such elimination seldom saves time in the long run." (Taylor, 1969)

4. Technologists want pilot plant studies to get design data for scale-up. Product is secondary. Marketing personnel see the need for product for test market or consumer research studies as paramount. These are the seeds for conflict.

5. Engineers tend to specify unique process designs based on the available data. Such designs require unique purpose-built equipment, which is expensive. Over the life cycle of the product, costs may not be recovered fully, or it may be a total loss if the development project is a failure. Consequently, standard, off-the-shelf equipment is used, and scale-up efficiencies are thereby compromised.

6. Echoing 1, 5 and to an extent 2, the state of process control in the food industry is still not highly developed. Off-line controls and batch operations are the norm, with on-line controls a close second, while in-line controls delivering data in real time are only beginning to be developed to their full potential.

7. "The present food industry is still fairly labour intensive and in some parts of the country the labour force is unskilled with a high turnover rate." (Taylor, 1969)

Taylor's remarks, made more than 20 years ago and intended for the U.K., apply equally well today in North America and in most major food processing areas.

If no recipe exists for technologists to begin with, one will have to be developed. Analysis of similar competitive products already in the marketplace provides clues to recipes. Many products are fairly crudely broken down into their component ingredients with such simple procedures as sieving, filtering, and microscopic analysis. For example, plant particulate matter is readily identified through sieving: counts of pieces (or their weights) give estimates of the quantities of each used. Microscopic analysis will provide information on starches used, may provide identification of pureed plant pulp, and may even help identify spices and herbs.

In addition, ingredient lists on labels, cost comparisons of the competitor's product combined with a knowledge of current ingredient costs, competent objective sensory evaluations, and some simple chemical analyses will reveal much about a competitor's formulation.

More sophisticated analytical procedures — and there are any number of commercial laboratories that can do these — may be required to break down other more complex products to reveal ingredients. Flavor houses can supply information about flavors and are able to duplicate most flavors presented to them.

With basic recipes in hand, food technologists can now begin to experiment with ingredients to obtain the desired flavor, texture, acidity, saltiness, color, or whatever particular attributes are required for the final product. Also, work can begin on the stabilizing system to be used to obtain a safe product of acceptable quality throughout its expected shelf life.

B. Food Safety Concerns

Two major questions need to be answered when prototype recipes or products have been obtained:

- First, what hazards of a public health significance are or could be associated with these products? with their storage? with their use?
- Second, what desired quality attributes are to be built into these products that require stabilizing or maximizing?

These beg other ancillary questions:

- What are the major spoilage routes of products of this particular nature and composition?
- What duration of acceptable quality shelf life is desired for these products?

These desired attributes could be organoleptic attributes, specific dietary or nutritional attributes, functional attributes, or attributes for convenience of preparation.

Elements of quality and safety must be designed into products at the inception of development. This is, in effect, the beginning of a hazard analysis critical control point (HACCP) program. It begins with the product design.

Concern over the safety of any food product, whether it be an established product or a new product, centers around all of the following:

1. presence of food intoxicating microorganisms exceeding recognized norms in the product itself or in the ingredients and raw materials forming part of the product which, when ingested by susceptible consumers, may cause illness or death,
2. presence of toxicants of a biological source, which may enter with the ingredients or be formed during processing. These could be preformed enterotoxins or mycotoxins or compounds such as domoic acid of algal blooms,
3. presence of chemical hazards in the food product or in any of the ingredients, which may occur with the carry-over of pesticides, herbi-

cides, growth stimulants, or even fertilizer uptake, or the appearance of these, which may arise from processing through intercompound reactions;

4. presence of miscellaneous extraneous matter, such as stones, glass fragments, metal pieces, or wood, in the food product, which can cause serious injury if ingested,
5. presence of insects and insect parts, an esthetic hazard, which can have a shock effect on consumers, resulting in illness.

Items 3, 4, and 5 should be prevented from entering the food or removed from the food by a total quality management program (Shapton and Shapton, 1991).

Many processing steps (trimming, cleaning, and blanching), plus plant support systems (good manufacturing practices (GMP), hazard analysis critical control point programs (HACCP), cleaning, and sanitation), are essential in reducing the numbers of microorganisms jeopardizing the safety and stability of products (see, for example, Shapton and Shapton). For products that are to be minimally processed, extreme attention to a total quality management program is essential.

The concept of safety is evolving. Its interpretation is broadening and coming under closer scrutiny from regulatory agencies in all countries.

Safety concerns have centered on susceptible consumers, that is, the very young and the elderly. But the concept of the "susceptible consumer" must be expanded now to include immuno- or health-compromised individuals. Brackett (1992) describes these as

> people with underlying chronic health problems such as cancer, diabetes, or heart disease; individuals taking certain immunocompromising drugs, such as corticosteroids; and individuals with immune deficiency diseases such as acquired immunodeficiency syndrome (AIDS).

The latter are becoming a larger proportion of the consuming public. They are more susceptible than normal individuals to food intoxicating microorganisms: they may fall ill after ingesting much smaller numbers of microorganisms than would harm normal consumers. Archer (1988) reports that AIDS-infected male patients are 300 times more susceptible to listeriosis than are AIDS-negative males.

This new "susceptible consumer" is one that technologists must consider in designing safe stabilizing systems for new products.

Emerging recognition by food microbiologists of the ability of some exotic and well-known microorganisms to grow and become health hazards in stored chilled foods is reshaping the thinking of developers of new food products. Indeed, as the limits to growth and viability of food-associated microorganisms in multiparameter stabilizing systems (hurdle technology; see Chapter 4, Section H) are studied, some anomalies of accepted limits of growth and viability are appearing. The importance of rigid adherence to stabilizing factors has grown.

Pathogenic psychrotrophs capable of growth around 5°C are *Listeria monocytogenes, Yersinia enterocolitica*, type E *Clostridium botulinum, Vibrio parahemolyticus, Vibrio cholerae, Bacillus cereus, Aeromonas hydrophila, Staphylococcus aureus*, enterotoxigenic *Escherichia coli*, and some *Salmonella* species (Farber, 1989). Any temperature abuse of susceptible products in the chain from processor to consumer could be potentially dangerous.

Konuma and colleagues (1988) studied the incidence of *B. cereus* in meat products in Japan. They determined that the most likely source of contamination was from meat product additives and not the meat itself. Beckers (1988) reviewed the incidence of foodborne diseases in the Netherlands (1979 to 1982) and found *Bacillus cereus, Salmonella* sp., *Campylobacter jejuni, Staphylococcus aureus, Clostridium perfringens*, and *Yersinia enterocolitica* to be major causative microorganisms. Meat and meat products, fish and shellfish, snacks, and foods prepared for immediate consumption, such as in food service outlets, especially those catering to hospitals, old peoples' homes and cafeterias and restaurants, were the primary carriers. Hemorrhagic colitis associated with *E. coli* has been frequently reported in nursing homes (Stavric and Speirs, 1989). Processed packaged meat products were found by Tiwari and Aldenrath (1990) to be contaminated by *Listeria monocytogenes*. Slade (1992), as a result of an extensive review of *listeria* in food processing environments, considers the presence of this microorganism to be ubiquitous.

In developing stabilizing systems for food products, developers must be watchful that safety is not or could not be compromised. This may not be as simple as one might think. Interactive packaging, modified atmosphere packaging, and controlled atmosphere packaging (MA/CAP) utilize gas mixtures inside packages that can control the growth of some spoilage microorganisms but certainly not all microorganisms. Brackett (1992) noted that spoilage was retarded in MAP produce but *A. hydrophila* and *L. monocytogenes* were unaffected by the modified atmosphere in the package. That is, visual spoilage was stopped, but toxic microorganisms could grow. Idziak (1993) explained to me that some gas mixtures used to extend the shelf life of many chilled products are the very ones used to culture anaerobic pathogens.

If chilled storage is compromised, growth of anaerobes may proceed with no visual indication of spoilage being apparent. Park and co-workers (1988) confirmed this in a study of MA-packaged commercially processed wet pasta. The gas mixture was $CO_2:N_2$; 20:80. Packages were stored at good (5°C) and poor (16°C) temperatures. The latter temperature is frequently encountered in open refrigerated shelves in retail outlets. The products were stated to have a recommended shelf life of 4 weeks. At the end of this time, sufficient staphylococcal enterotoxin had formed to have caused illness in "sensitive individuals" in a significant portion of those samples held at the poor refrigerated conditions.

Many products, e.g., anchovies, unpasteurized acidified foods, beef jerkies, are "semiconserved". That is, they are safe, stable, wholesome products in their stabilizing system; for anchovies, this is the very high salt-on-water they contain. By themselves, anchovies can be eaten safely: they are a preserved product. When, however, they are used as ingredients in another product, e.g., an all-dressed pizza, a quiche, or a flan, in which the safety system of the semiconserve is compromised, the potential for a food hazard to develop is very real. If the pizza or quiche or flan is held for a long period of time in a warming oven or in a poorly maintained refrigerator, the microorganisms in the anchovies (to continue this example), some of which are spore formers held in bacteriostasis by the high salt content of the original anchovies, are no longer in such inimical conditions. Growth can commence, and spoilage or intoxication can occur.

Both Archer (1988) and Rowley (1984) draw attention to the importance and unusual behavior of sublethally injured pathogenic and spoilage bacteria during processing. Sublethal injury to microorganisms, whatever the nature of the injury, evokes an adaptive response. In some instances, "injured" microorganisms may become more resistant (Archer, 1988). Pretreatment, i.e., processing, of foods may alter the characteristics (sensitivity) of microorganisms to stresses.

The desire for less highly processed foods, such as chilled foods, opens up a Pandora's box of new problems concerning safety and stability that technologists have never been forced to deal with before. These new safety concerns must be addressed by development technologists with the quality control and processing departments. Together they must institute suitable good manufacturing programs (GMP) and hazard analysis critical control programs (HACCP) for products under development to control hazards adequately.

C. Food Spoilage Concerns

Margaret Hungerford's well-known comment: "Beauty is in the eye of the beholder" could very easily be rewritten by food technologists, "Spoilage is in the eye of the beholder". The beholder, for which read "the consumer", decides the acceptability or spoilage. Any definition of spoilage fails if it does not respect the consumer. For consumers, spoilage is the failure of products to meet the expectations (or the consumers' perception of those expectations) promised by developers. The product has lost its appeal and is rejected. This is not suggesting in any manner that spoilage is a marketing problem. Spoilage simply leads the consumer to think, "You promised me this, but the product I bought gave me ...." Developers can substitute "bad smelling", "bad textured", "unsightly", "moldy", etc. into this sentence.

Products must be safe for ethical and humanitarian reasons: products must be stable, i.e., unspoiled, for economic and esthetic reasons. Safety concerns for new products are far more important than spoilage concerns. Nevertheless, spoilage of products in the marketplace is a serious problem.

Especially for products in a test market, spoilage represents an enormous loss in research and development time and money, a loss of promotional money, a great blow to the company's brand image, and a loss of consumers. The impact of a defective product in the marketplace should be remembered (Ross, 1980): 1 out of 25 consumers who encounter a defective product complain, but among those who did not bother to complain, there were at least six who had a serious complaint (see Chapter 2). As if that were not enough damage to a product's image, on average, each complainant tells nine others. This represents many disgruntled consumers.

The new food product developer must determine how to stabilize the food to provide, in addition to safety, the maximal quality attributes throughout the anticipated shelf life. Safety and spoilage are allied problems, which are treated similarly.

Spoilage in food products results in a loss of quality and can be ascribed to any, or all, of the three following causes:

- biological changes resulting from the activities of microorganisms present in foods, in ingredients added to foods, or in the environment in which foods were prepared,
- enzyme systems within foods which, if uncontrolled, cause autolytic or oxidative changes within foods,
- nonbiological causes — chemical and physical reactions that can take place within a food matrix between the various chemical species present as a result of processing, storage, or packaging.

These biological and nonbiological causes result in nutritive, textural, visual, and organoleptic changes in foods, some of which are summarized in Table 3.

Interactions between the package, the storage environment, and the food product itself can occur, particularly in distribution channels. Corrosion of aluminum containers by high-salt foods and detinning of metal containers by high-acid foods are well known to food technologists. Fluctuations in cold storage or freezer storage temperatures, which are frequently encountered in warehouses and in distribution channels, can cause a loss of esthetic appeal with the appearance of large ice crystals in frozen produce.

Certain aromatic flavors are not compatible with some plastics used either as the container itself or in the lining of the container. Likewise, some flavor chemicals can be quite reactive in the presence of metal packaging and can cause corrosion of the packaging and the development of off-flavors.

Not all of the changes noted in Table 3 are undesirable. Oxidation, for example, is desirable in alcoholic beverages in the aging of whiskey and in the development of color, but it is not at all desirable in fats and oils or in products with a high content of these. Food technologists must sort good reactions from bad and prevent or control those changes that are deleterious to the quality of the product or to its stability. Measures used to control or prevent the changes

TABLE 3. Categories of Changes in Foods With Examples of Such Changes

The Reaction Mechanism		
Physical	Chemical	Biological
Phase changes	Oxidation	Respiration
Evaporation	Reduction	Oxidation
Concentration	Hydrolysis	Autolysis
Crystallization	Condensation	Fermentation
Mass migration	Decarboxylation	Putrefaction
Irradiation	Deamination	
	Maillard reaction	
The Sensible Changes in Foods		
Exudation	Off-flavors	Wilting
Separation	Off-odors	Softening
Precipitation	Discoloration	Discoloration
Clumping	Browning	Slime formation
Clotting	Fading	Clotting
Textural changes	Exudation	Exudation
Grittiness	Textural changes	Off-flavors
Staling	Container interactions	Off-odors
Toughening	Rusting	Toxins
Discoloration	Delamination	Senescence
Fading		Excessive cfu/gm
Opacity		

vary in their severity. Consequently, treatments to stabilize food products may, in turn, cause changes in food products that marketing personnel will consider undesirable in the finished product. Or, costs of control, such as special packaging techniques, may add too much to the bottom line. The use of chemical additives, even though they are permitted, may be an unacceptable alternative if marketing personnel want to project a "natural" image for a product with a clean label.

Mossel and Ingram (1955) organize microbial spoilage of foods as follows:

- intrinsic factors that are a characteristic of the food, its processing, and its ingredients and additives; these would include its pH, a_w, antimicrobial constituents, colloidal state, or biological structure,
- extrinsic factors, which include all environmental factors, such as temperature, relative humidity, or gases, which may be in contact with the food or its package,
- implicit factors, which embrace the microorganisms involved, their population dynamics with all the synergistic and antagonistic pressures, and microbial metabolism; these can be influenced by both intrinsic factors (pH and antimicrobials) and extrinsic factors (temperature).

Such an organization of factors in microbial spoilage permits developers to consider stability from three broad approaches: the food itself and its formulation and processing, the distribution and packaging systems to protect the food, and the type and numbers of microorganisms present.

Clues regarding which stabilizing techniques to use to control biological and microbiological hazards will be found from

1. a knowledge of what microbial hazards are associated with the type of food under development or with the processing system to be used in preparing the ingredients,
2. the growth characteristics of microorganisms implicated with the microbial hazards. Microorganisms can be crudely classified according to their growth characteristics:
   * temperature of optimum growth (psychrophile, mesophile, and thermophile)
   * reducing environment (Eh)
   * optimum pH requirements for growth
   * water relationships (a_w)
   * specific nutrient requirements
   * sensitivity to competitive pressures of other microorganisms
   * sensitivity to the presence of certain chemical agents
3. a knowledge of the chemical and biological properties of the food and its ingredients.

These clues provide the technologist with suggestions for various techniques with which to provide the desired degree of safety and to stabilize the product.

A simple example will illustrate the principle. Consider that food technologists have determined that the expected course of microbial deterioration of a hypothetical new food product will be an equally hypothetical microorganism, which has the following characteristics:

1. grows poorly in pH conditions below 4.8,
2. does not tolerate water activity conditions below 0.9,
3. is a mesophile,
4. prefers aerobic conditions for growth,
5. is sensitive to sorbate as a preservative.

Food technologists have several avenues of stabilizing the shelf life that can be followed up. They can:

* acidify the product to below pH 4.8,
* lower the water activity of the product below 0.9,
* consider a chilled product with refrigerated distribution,
* employ vacuum packaging and/or increase the reducing potential of the food matrix and/or consider modified or controlled atmosphere packaging,

- use sorbic acid or its salts as a preservative,
- heat treat the product.

Therefore, there are six methods with which to provide some degree of stability to this hypothetical product. If combinations of techniques are considered, there are several more multiparameter methods to use. Not all the methods may be suitable for the product, nor will all be permitted to be used with the product or be acceptable to the company for use.

Enzymatic changes in foods are controlled similarly to the control of microorganisms. Heating to inactivate enzymes (blanching) is common. However, the nature of the final product may not allow food technologists to heat-inactivate enzymes. Technologists must then resort to other means in order to control unwanted enzymatic actions: e.g., chilling or freezing to slow the enzymatic reaction, substrate removal or use of substrate antagonists, or specific chemical inhibitors of the enzyme in question.

Chemical and physical changes that make foods less acceptable, thereby contributing to a shorter shelf life, are somewhat more complex to treat. The rate of a chemical reaction can be slowed by lowering the temperature. But by lowering the storage temperature, solubility of solutes may be affected. These can become concentrated to the point that they crystallize out. As temperature is lowered, phase changes can occur. Water becomes ice; oils solidify. Emulsions can be destabilized with extreme temperature changes.

Certain processing techniques can be used to delay or prevent chemical and physical changes:

- Homogenization will prevent separation of oils.
- Agglomeration and crystallization will prevent caking and improve solubility.
- Choice of ingredients (for example, crystallizing and noncrystallizing sugars) and additives (doctoring agents) will prevent staling (moist cookies) and crystallization (grittiness).
- Encapsulation will separate labile components from other reactants in foods and minimize flavor losses.

Some foods are quite stable, while others can spoil rapidly if precautions are not taken to preserve them.

The characteristic stability of foods is the basis for a classification of foods. The simplest classification according to stability uses three categories (McGinn, 1982):

1. highly perishable foods with a short period of acceptable quality shelf life. This is measured in days. Some examples are fluid milk, fresh meats (meat, poultry, and seafood), some fresh produce (leafy green vegetables and soft fruits), delicatessen salads, fresh bread, and specialty cream- or custard-filled baked goods.

2. semi-perishable foods with an acceptable quality shelf life calculated in weeks. Conserved meat products (bacon, hams, some fermented and semidry sausages), some bakery goods (fruit cakes), dairy products (natural cheeses), potato and tortilla chips, and other snack foods are typical examples of this category.
3. highly stable foods with an acceptable shelf life determined in months or years. Dehydrated foods and food mixes, canned foods, many cereal products (flour, pastas, breakfast cereals), confectionery products (toffees, hard candies, and some chocolate products), jams, and jellies typify this group.

It is, of course, assumed in the above classification that storage conditions proper for the type of food prevail and that sound packaging contains the product and protects it from any inimical external factors. Shelf life of any food product and its safety are, of course, jeopardized if adverse storage conditions exist or if container integrity is broken.

In general, as the shelf life of products lengthens, the causes of instability alter from biological (largely microbiological) to physical or chemical causes. For example, ground meat or pasteurized milk, both short shelf life products, spoil for microbiological reasons, not for chemical or physical reasons. On the other hand, UHT (ultra high temperature processed) milk has a long shelf life and is more likely to fail for flavor and textural (grittiness due to lactose crystallization) reasons. Beef jerky, a product with a long shelf life, may fail for reasons of off-flavors due to fat oxidation. As the duration of the shelf life increases, biological factors, while always important, are overshadowed by chemical and physical changes in the food brought about by temperature changes, light, presence of oxygen, and abusive handling. These, in turn, cause changes in flavor, texture, color, and odor. Finally, with long-life, stable products, biological concerns as factors in spoilage are minimal but always present. For example, McWeeny (1980) has pointed out that in the long-term storage of flour, lipases and lipoxidases can cause losses of "oven spring". However, with long-life products, spoilage is usually due to chemical and physical changes in the food, causing extensive sensory alterations in the product.

The above classification is not hard and fast. The shelf life of any food will depend on the processing treatment it receives. In the dairy category (generally considered a short shelf life category) some hard dry cheeses and in the meat category (also short shelf life products) some dry sausages have shelf lives of months or even years.

The expected duration of quality shelf life, described in the product statement, will determine which preservative techniques must be used to sustain the product through warehousing and distribution channels to the marketplace and to consumers' tables. Stabilizing systems for products with an expected shelf life of 3 months will be different from those same products if marketing personnel decides, unrealistically, that they are expected to have a

2-year acceptable quality life. Ingredients, processing, packaging, storage, and distribution will be quite different for the two products.

The various quality attributes of any food product will deteriorate, but each attribute may deteriorate at a different rate with respect, for example, to changes in temperature. That is, the mechanisms for spoilage have different rates of reaction; they have different Q_{10} values (Labuza and Riboh, 1982; Labuza and Schmidl, 1985). At one temperature, the color of a beverage, for example, may alter faster than its flavor. At another temperature, off-flavors may proceed more quickly. In addition to temperature, the causative mechanism could be light. Color in beverages, for example, can be stabilized with a colored glass bottle or with the use of an opaque foil/paperboard-laminated container. However, the latter is considered environmentally unfriendly and is even banned in some areas.

Tortilla chips can spoil by losing their crispness or by going rancid. At one manufacturer of tortilla chips, technologists, on the basis of in-plant taste testing, considered that rancidity was the limiting factor in the desired shelf life of an up-scale tortilla chip product. Accordingly, the product was packed using a gas flush of nitrogen. Rancidity was stopped, but loss of crispness became the limiting factor. As well, surprisingly, consumer complaints regarding flavor poured in. Consumers liked the hint of rancidity in the product. With that gone, they stopped purchases, leaving the product sitting longer on the shelves and losing crispness.

Spoilage is in the eye of the beholder. Technologists have many questions to resolve. What quality features are lost? What shelf life expectations are required? Stability is being measured with respect to which quality attributes of the product? Equally important to food technologists is how those quality features are being lost or altered. Answers to these questions demand that technologists find stabilizing systems to give the desired shelf life as well as the degree of safety from hazards of public health significance that consumers have the right to expect.

III. MEETING THE CHALLENGE: NEWER FOOD STABILIZING SYSTEMS

For simplicity, the expression "stabilizing system" will be used for processing technologies embodying mechanisms that ensure safety of foods with respect to hazards of public health significance and their quality throughout their expected shelf life. Both safety and quality are important considerations in the development of new food products. Technologists must be aware that hazards to spoilage and safety may have different mechanisms in the food products under development.

To stabilize foods, technologists have numerous techniques that can be used (Table 4). The techniques can be used singly or in very deliberately

TABLE 4. Techniques to Stabilize Foods

Stabilizing Stress	Possible Mechanisms
Thermal processing	
Temperature > 100°C	Spore inactivation
Temperature < 100°C	Vegetative cell destruction
	Enzyme inactivation
Chilling	
Refrigeration	Slowing of microbial metabolic pathways
	Slowing of chemical reactions
Freezing	Immobilization of water
	(some microbial destruction)
Fermentation	Alteration/removal of a substrate
	Acidification
	Production of antimicrobial agents
	Overgrowth by beneficial or benign microorganisms
Control of water	(Partial) removal of water
	Humectants (control water activity)
	(Freezing)
Acidification	Hostile pH for microorganisms
	Suboptimum pH for enzymic reactions
	Preservative (specific property of acid)
	(Fermentation)
Chemicals	Specific preservative action
	Modification of a substrate
	Enzyme antagonist
	Specific chemical action (antioxidant)
	Control metabolic pathway
	(Control redox potential)
Control of redox potential	Prevents senescence
	Prevents growth of some harmful microorganisms
Irradiation	Inactivates microorganisms
	Inactivates stages in metabolism
Pressure	Protein denaturation
	Disruption of cell organization

designed combinations, planned uniquely for a particular product. Each technique confers, according to the nature of the food, varying degrees of stability through different mechanisms (Table 4). Knowledge of these mechanisms for bestowing stability will dovetail with food technologists' knowledge of the mechanisms for spoilage and for safety, to allow the design of the most suitable stability system for the product.

Yet, numerous as these techniques are, there are limits to what can be done by technologists. Some stabilizing systems may be unacceptable because of the damage they do to the defined nutrient standard of a product. Some may alter the textural or flavor characteristics that marketing people wish particularly to emphasize in their promotions. Thermal processing, freezing, or acidification can produce significant changes in both flavor and texture. Even with the most advanced technology, some products will not be successfully stabilized without severe alterations in their character.

There are also nontechnological constraints to what stabilizing systems are available to technologists. Freezing, for example, may well be impossible as a procedure for preservation if a company's manufacturing, marketing, and distribution capabilities cannot handle such a product. Both the use of chemical preservatives and irradiation may be considered to be techniques with too much consumer resistance. The use of certain packaging materials may be restricted by legislation or may require the establishment of a collection system for recycling (Akre, 1991) or may simply be unpopular with environmental activists. Marketing personnel's or senior management's reluctance to use controversial or sensitive procedures acts, in effect, like a screen.

Both new developments for stabilizing food products and improvements in older, traditional technologies have emerged (Best, 1988). Application of these to food product development has permitted exciting new products to find new market niches and even to supplant older, established products. Some of these technologies and techniques will be reviewed here, not because they are the newest or most novel but because they show promise in providing consumers with products they seem to want.

A. Thermal Processing of Foods

In thermal processing, the more rapidly heat can be transferred into a food and the more rapidly it can subsequently be removed, the less heat damage there is to the quality attributes of the food. This has long been clearly understood by technologists. Getting heat into and out of food rapidly was the answer to the problem of producing high quality canned foods. Higher temperatures could be used to inactivate heat resistant microorganisms or spores, and therefore shorter processing times could be had, but only if the thermal conductivity of the food and its thermal diffusivity permitted the rapid movement of heat into and out of the product.

With the use of continuous flow heat exchangers, HTST (high temperature short time) or UHT (ultra high temperature) processes, followed by cooling and then aseptic filling, can be successfully used on many foods. The problem is, of course, that food containing a considerable proportion of large-sized particulate matter is not adequately heat processed. The required residence time in the heat exchanger for the more fluid convection heating portion of the food is much less than for the thicker, conduction heating portion of the food.

It has long been recognized that thermal processes for foods could be drastically reduced if heat penetration could be speeded up by agitating the food while in its container in the retort. Many commercial retorts now can agitate cans in an end-over-end fashion, axially, or in a rocking motion. As the container is rotated in the retort, the headspace in the container moves through the food, mixing the contents and thereby assisting heat penetration to the cold spot. Thus, a shorter thermal process is obtained with improved product quality. With the faster rate of heat transfer, higher processing temperatures can be used with still shorter process times (HTST and UHT processes).

In general, agitating retorts employing end-over-end can rotation or a rocking motion are batch type retorts. Retorts that provide axial rotation to agitate the container's contents are batch or continuous-type retorts.

If the geometry of the cylindrical can were to be altered to permit more rapid penetration of heat, heat damage to the contents of the container could be minimized still further. Thin profile (or low profile) containers, such as the (flexible) retort pouch, the semirigid container, or the larger institutional half-steam table tray, have an altered geometry. They present two broad surfaces for heat penetration. Consequently, heat penetration, whether by conduction or convection, is more rapid in thin profile containers than in conventional round, cylindrical containers (Rizvi and Acton, 1982; Tung et al., 1975; Chapman and McKernan, 1963). Rizvi and Acton (1982) demonstrated more rapid heat penetration and cooling with the thin profile container (in this case a pouch) over the conventional can, using equal volumes of sweet potato puree processed to the same sterilizing value. The broad, flat surfaces of the thin profile container plus the shallow depth of product permit very rapid heating and cooling. Cooling started in the pouch after approximately 30 min, whereas cooling of the can commenced after approximately 80 min. The can contents were subjected to more heat damage.

Again, Tung and colleagues (1975) accomplished a thermal cook (average $F_0 = 8.2$ at 121°C) in 33 min for 14 fl oz of cream style corn still-processed in a flexible pouch and in 75 min in a 300 × 407 tinplate can. As a consequence of the more rapid heat penetration, heat damage to the food contents was minimized. Quality of the product with respect to color, flavor, nutrition, and integrity of particulate pieces was greatly improved.

The retort pouch has become a very popular container in Japan for thermally processed sauce-based products, such as curries, spaghetti sauces, or stews (Saito, 1983), and it is certainly well known in military rations (Lingle, 1989; Tuomy and Young, 1982; Mermelstein, 1978). It has not been received with any great success in the consumer market in North America despite its many advantages (Mermelstein, 1978).

Developments in these two areas then, thin profile containers and agitating retorts have given thermal processing a new appeal for the production of added value, high quality gourmet products. Many of these are gaining acceptance as main course items, particularly in the food service industry (Eisner, 1988; Adams et al., 1983).

B. Ohmic Heating

Ohmic heating overcomes many of the problems encountered with continuous flow through plate heat exchangers. In ohmic heating, heat is developed within the food itself by passing a low frequency alternating current through the conducting food product. Heating is not primarily dependent on either convection or conduction. Heat penetration is largely dependent on the electrical conductivity of the food. Consequently, heat penetration is less affected by the presence of particulate food, its size, or other physical properties such as thermal diffusivity (Biss et al., 1989; Selman, 1991; Halden et al., 1990). The result is that there are not large temperature gradients between the more liquid portion of the food and particulates in the food, since these heat at roughly the same time. Packaging is, of course, done aseptically. The development of ohmic heating is fully described by Biss and co-workers (1989).

In addition, since there are no heat transfer surfaces, as in plate heat exchangers, there are no burn-on deposits to foul the heat exchange. Energy conversion is rated at "… greater than 90% compared with only 45–50% for alternative forms" (Selman, 1991).

Ohmic heating does require that the product be an electrical conductor. For most foods with water contents in the 30 to 40% range and dissolved ionic constituents, this requirement does prevail. However, nonionized food components, fats and oils, sugar syrups, alcohols, and nonconducting solids, such as bone and cellulosic material, can be heated only indirectly by ohmic heating.

Halden and co-workers (1990), studying the changes in the electrical conductivity of foods during ohmic heating, observed the following:

- The electrical conductivity of pork meat, chicken, and other meats varied only slightly with temperature.
- Pork meat and pork fat, if ohmically heated together, would heat at different rates. Therefore, during processing, thermal equilibrium may not be attained through conduction, and a process must be designed to sufficiently heat treat the slowest-heating food component.
- Aubergines heated ohmically broke down structurally. Large holes were formed in the tissue. Strawberries softened markedly around 65°C. This suggests that there may be very specific tissue responses to ohmic heating.
- Starch gelatinization caused a change in conductivity.
- Preheating increased the electrical conductivity of some foods.

They caution that "electrical conductivity data from sources other than ohmic heating must thus be treated with care when designing an ohmic process".

Ohmic heating as a stabilizing system is yet in its infancy, but it does show promise as a system to rival plate heat exchangers for rapidly heating multicomponent foods containing nonuniform particulate material and liquid. Heat damage to sensitive quality characteristics could be minimized.

C. Stabilizing with High Pressure

The use of high pressure to stabilize food systems is receiving attention as a novel technique to process foods. The history of and developments in the technique have been reviewed by Farr (1990), Hoover et al. (1989), and Hayashi (1989). Surprisingly, the technique is nearly 100 years old and was first reported by Hite in 1899 for the preservation of milk (reported in Hoover et al., 1989).

The pressures that have been used are in the order of 3500 atm to nearly 10,000 atm (1 atm = 14.696 lb/in^2 = 1.033 kgm/cm^2). Changes occur in proteins and therefore in protein foods and in food microorganisms from 1000 atm and up. For example, Okamoto and co-workers (1990) produced hard gels of both egg yolks and egg whites that stood under their own weight with applied pressures of 4500 kg/cm^2 and 6000 kg/cm^2, respectively, at 25°C for 30 min. Softness and adhesiveness of these gels could be varied with the amount of pressure applied. As the pressure was increased, the hardness of the gels increased; their cohesiveness increased, but adhesiveness decreased. A solution of soy protein pressurized at 2000 kg/cm^2 produced a soft gel, which deformed readily, whereas, at 4000 kg/cm the gel was hardest. Nevertheless, pressure-produced gels were softer than heat-produced gels. Differences between heat-produced gels and pressure-produced gels led Okamoto (1990) to suggest that the mechanism of gelation was different in the two techniques. This qualitative difference is a factor that developers must consider, should this technique be used. Taste and color were unchanged in the pressurized foods.

Surimi from different fish sources is gelled with high hydrostatic pressure in Japan. Also in Japan, jams, with their high acidity and solids content, are further preserved by high pressure (Farr, 1990).

Metrick and her colleagues (1989) studied the sensitivity to high pressures (2380 to 3400 atm) of *Salmonella senftenberg* and *Salmonella typhimurium* in both a phosphate buffer and strained chicken baby food. Cell death, as well as cell injury, occurred in this pressure range and increased with increasing pressure. Inactivation was greater in buffer than in the chicken medium, suggesting that the medium may have a profound effect on survival of microorganisms. Again, this is a factor developers must consider in designing safe systems. The more heat resistant strain of *S. senftenberg* was more pressure sensitive than was *S. typhimurium*.

Hayashi (1989) cites the advantages of high pressure to stabilize foods as the avoidance of heat damage and the preservation of natural flavor, taste, and nutrients. Hydrostatic pressure is transmitted instantaneously and uniformly into food, unlike during heat transmission, in which thermal conducting properties influence the rate of transfer. It can be used to kill insects as well. High pressure can also be used in conjunction with other stabilizing methods, such as acidification, antimicrobial agents, or mild heat, to stabilize foods. Pressure appears to be more efficient in inactivating cells in acid pH than in neutral pH (reported in Hoover et al., 1989). High pressure equipment capable of pressures up to 130,500 psi and up to 80°C is available from ABB Autoclave Systems, Inc. (Columbus, OH).

Selman (1992) provides further information on costs of the process.

D. Other Nonthermal Stabilizing Systems

A summary of nonthermal processes for stabilizing food systems is given by Mertens and Knorr (1992). Among these less-damaging processes they list the following:

- high-electric field pulses to cause dielectric cell membrane rupture. They cite two industrial applications, one to improve fat recovery from animal slurries from slaughterhouses, and the other to stabilize pumpable foods;
- oscillating magnetic field pulses: "When a large number of magnetic dipoles are present in one molecule, enough energy can be transferred to the molecule to break a covalent bond ... certain critical molecules in a microorganism, like DNA or proteins, could be broken by the treatment, hence destroying the microorganism or at least rendering it reproductively inactive" (Mertens and Knorr, 1992). No commercial details were available;
- microwave effects, in which the authors briefly and concisely review the nonthermal effects of microwaves. Although they leave the impression that they are in agreement with others that the inactivation of microorganisms is primarily the result of the thermal effects of microwaves, they do state: "If we assume (deleterious cellular effects) are real, it is difficult to imagine how these sub-lethal and long-term effects can be upgraded to a useful food-preservation method";
- linear induction electron accelerator to produce ionizing radiation in place of using radionuclides;
- intense light pulses, in which the inactivation effect may be a combination of photochemical effects and photothermal effects;
- carbon dioxide treatments, whose effects are due to a combination of a pH drop within the cellular structure of the food and displacement of oxygen;
- the use of chitosan (deacylated chitin) as an antifungal agent (see later);
- antimicrobial enzymes (for example, the use of enzymes to remove a necessary metabolite from a food);
- biological control systems that could be used in conjunction with the hurdle concept of food preservation.

The advantage of all these techniques is, of course, the minimization of the sometimes undesirable changes that thermal processing brings about in foods.

E. Control of Water

Water usually plays a role in all spoilage reactions. Water is necessary for enzymic reactions. Water brings chemically reactive species together. It also brings nutrients to microorganisms and permits the movement of motile species. As awareness has grown of how the control of water in foods could stabilize those foods, the concept of water activity (a_w) has developed, making possible a whole new range of foods, including semimoist foods or intermediate moisture foods (IMF). Early developments in IMF can be found in Davies et al. (1976), which is still one of the best basic books on the subject.

As important as a_w is in the stabilizing of a food system, water's influence elsewhere in foods must also be curbed. Water activity itself is a comparatively new concept to many in food manufacturing, but it is as Best (1992) has aptly described it, a minimal concept that is undergoing a major upheaval.

This upheaval has largely been instigated by Slade and Levine (1991), whose article has caused a revision in thinking on the concept of a_w. They introduced the consideration of phase transitions in foods. This topic is taken up again under the topic of sugars in a later section on ingredients.

Roser (1991) reviews a novel use of the sugar trehalose in the drying of foods. Trehalose is found in high concentrations in so-called cryptobiotic organisms. Cryptobionts have a remarkable ability to survive harsh treatment or conditions. In his review, Roser cites the "resurrection plant", which, when fully dry, can withstand heating to 100°C and megarad doses of irradiation. The secret would appear to be the ability of trehalose (and other compounds with this ability) to dry as glasses rather than as crystals. Trehalose shows promise as an aid in producing superior dehydrated products.

Jezek and Smyrl (1980) describe the dehydration of apple slices by immersing the slices in a 65% w/w sucrose solution to remove water, followed by further drying by vacuum. They found an increase in volatiles and minimal heat damage to a low heat treatment.

F. Controlled Atmosphere/Modified Atmosphere Packaging (CA/MAP)

Gases, used either singly or in combinations of precisely defined mixtures, have given a longer fresh shelf life to fruits and vegetables. These same gases and mixtures of gases can be used to control the metabolic pathways of fresh fruits and vegetables, both in bulk storage and in unit packages; they can also extend the shelf life of many other added-value products.

This has given rise to controlled-atmosphere and modified-atmosphere unit packaging (CA/MAP) to prolong the shelf life of many food products besides fruits and vegetables, such as meats. An overview of CA/MAP for fresh produce in Western Europe is provided by Day (1990).

Despite its many advantages, CA/MAP has not been as popular in North America as it has been in Europe. Day (1990) explores several reasons why this form of packaging has found greater acceptance in Europe than in North America. Product developers should be cognizant of these if they wish to apply the technique to products in North America. These reasons are

- Distribution is vastly different in the U.S. than in the European Economic Community (EEC). The U.S. has four times the land area as the EEC but

only three-fourths of the population of the EEC. Distribution requirements are more concentrated in the EEC.
- North Americans shop less frequently than European food shoppers.
- Chilled foods (which receive the most value from CA/MAP) do not have the acceptance in North America that they do in the EEC.
- CA/MAP technology and the quality systems to support CA/MAP are in place in the EEC but are not in North America. (Compare the distribution systems.)
- Higher food prices in the EEC have made CA/MAP products more acceptable economically than in the North American market, where prices are generally lower.
- In the U.K. and France, giant retail food chains dominate the marketplace, whereas in the U.S. there is more fragmentation (see the first reason above). These giants are determined in their advocacy of quality chilled foods.
- Different forces drive the markets in the EEC and North America. Retail forces drive the EEC marketplace, while in North America, it is driven by packer/consumer forces.

Day's (1990) first three points probably in large measure explain the lag in MAP technology and support systems in North America.

Anthony (1989) echoes some of Day's points when he suggests that the greatest strength for CA/MAP might be for premium, added-value products that are packed in this manner. However, he cautioned that the element of product liability is a factor, too. Any abuse in the chilled food CA/MAP chain can result in serious product losses and the potential for risks of public health significance.

CA/MAP stabilizing systems are generally used in combination with other preservative techniques, for example, refrigeration.

Quality in CA/MAP products can be compromised by abusive treatment throughout the chain, from producer to consumer to consumption. Abuse in CA/MAP chilled foods is a reality for product developers. For example, Slight (1980), in a study of U.K. storage and transportation of chilled foods, found that the inspection and maintenance of refrigeration systems for chilled foods was poor. In his survey, none of the transport refrigeration systems studied were operating properly. Again, while studying shelf stability of a new chilled Mexican entree item in southern California, I found that chilled food display counters in supermarkets could at times reach a high of 9°C (48°F). Moreover, they could hold at this high temperature for extensive periods of time. In still another example, Light and co-workers (1987) studied temperature distribution in chilled food vending machines in the U.K. They noted that some machines had ranges of fluctuation from −1 to +10°C. In addition, they discovered that some machines could not maintain the desired temperature range of 0 to 5°C for even 50% of their working cycle.

No wonder that Day (1990) suggests that distribution, as well as shopping habits and price structure, may all contribute to CA/MAP's slow acceptance in North America.

Both Day (1990) and Anthony (1989) caution that CA/MAP is no panacea for increasing shelf life. Both stress the need to control implicit factors, such as the numbers of microorganisms present, extrinsic factors, including the gaseous atmosphere, storage and distribution temperature, and humidity in the package, as well as the film composition, and intrinsic factors characteristic of the food itself. Geeson and co-workers (1987) found that CA/MAP successfully extended the shelf life of some varieties of apples but, in another paper, was unsuccessful in extending the shelf life of Conference pears (Geeson et al., 1991). The important point for developers is that one cannot take a preservative system successful for one product and apply it helter-skelter to another.

G. Irradiation

> We cannot control atomic energy to an extent which would be of any value commercially, and I believe we are not likely ever to be able to do so.

That excerpt is from a speech by Rutherford, considered by many as the father of the study of the atom, delivered to the British Association for the Advancement of Science in 1933. Rutherford's remarks notwithstanding, Lieber (1905) had earlier taken out a patent for the preservation of

> ... canned foods, meat, beef extracts, and other manufactured or prepared foods, milk, cheese cream, and the compounds thereof, fruits, jams, juices, jellies, and preserves generally....

He preserved the foods by impregnating them with emanations from "thorium oxid", thus rendering the substances radioactive! Irradiation has come a long way since 1905 and 1933.

A joint statement on food irradiation (AIC/CIFST, 1989) by the Agricultural Institute of Canada and the Canadian Institute of Food Science and Technology listed the benefits of irradiation over other more conventional processes as follows:

- Thermal degradation of foods is minimized, since irradiation can be applied while the food is frozen. (Irradiation is a nonthermal technique for stabilizing food — the food is not cooked as it is with microwaves.)
- Foods can be unitized, i.e., prepackaged, before irradiation, and thereby, contamination in distribution channels is minimized.
- The form of the food (liquid, solid, or semisolid), its shape, or whether it is unitized or in bulk is immaterial for irradiation.

- Irradiation can replace chemicals for preservative effects, decontamination, and disinfestation.
- Irradiation will prove useful in maintaining certain markets for pork (checking trichinosis), for poultry (salmonella control), and fruit (insect pests).
- Through irradiation's ability to provide a longer, pest-free shelf life, a greater variety of improved quality tropical products will become available in the marketplaces of more northerly climates.

Canada early gave approval to irradiation and has used irradiation since 1960 for sprout control in potatoes and onions and since 1965 for insect control in wheat.

Josephson (1984), long a proponent of irradiation, studied the advantages of irradiation for the military and concluded the following:

- Irradiation has the potential to provide shelf stable rations that are palatable, nutritious, and wholesome, yet with all the characteristics of fresh foods prepared for eating.
- Irradiated foods provide convenience and versatility, provide both meal components as well as snacks, and reduce preparation and labor in the kitchen.
- High quality, wholesome foods can be provided for far-distant military operations through irradiation.
- Irradiation is the ideal substitute for many food chemical preservatives, previously added to foods, which have proven to be "toxic, carcinogenic, mutagenic, or teratogenic."
- Irradiation is competitive, in respect to cost and energy consumption, with other conventional techniques for food preservation.
- Irradiation provides new opportunities in military feeding systems in all geographic and environmental conditions, whether in peace time or combat duties, and resolves many logistical problems encountered with conventional feeding systems.

Despite this enthusiastic endorsement, irradiation has been very slow to gain widespread acceptance in North America.

The effectiveness of irradiation (at the levels used in food processing) in inactivating enzymes and reducing the counts of microorganisms is largely due to reactive species such as hydrogen, hydrogen peroxide, and hydroperoxy radicals produced during the process (Robinson, 1985). Such reactive species are not unique to irradiated food. They can also be generated in nonirradiated foods (Robinson, 1985). The effect irradiation has on plant or animal tissue is dose dependent (AIC/CIFST, 1989; Gaunt, 1985; Giddings, 1984):

- sprout inhibition of 0.15 to 0.2 kGy,
- flour disinfestation of up to 1 kGy,

- spice cleaning of up to 5 kGy for 2 to 3 log cycles of reduction in count,
- parasite elimination of up to 6 kGy,
- reduction of bacteria of up to 5 kGy, depending on reduction desired,
- nonsporing pathogens of up to 10 kGy.

Irradiation has been used for several years to control infestations in fruit and vegetables in place of fumigation, to sterilize sewage sludge for conversion into fertilizer, and to sterilize surgical equipment (Anon., 1981). Tsuji (1983) describes in some detail the use of low-dose cobalt 60 irradiation to reduce microbial counts in fish-protein concentrate used as flavoring in the formulation of vitamin/mineral supplements in animal health products. Irradiation proved superior to gas sterilants (ethylene and propylene oxides). Tsuji estimates the cost of irradiation and associated handling at $0.07/lb.

A joint expert committee, under the auspices of three agencies of the United Nations, agreed that irradiation of food at doses up to 10 kGy presents no hazards of public health significance toxicologically, nutritionally, or microbiologically (AIC/CIFST, 1989). A steadily growing number of countries have accepted irradiation as a food process (Loaharanu, 1989). Consequently, any developer of new food products for the export market should be aware of local regulations of the importing country.

Food is irradiated by one of two means. Either gamma rays from a radionuclide, such as cobalt 60, or X-rays from a machine-made source are used to bombard the food. At comparable energy levels, gamma and X-rays are identical. Energy sources to produce the radiation beam are housed in thickly walled concrete bunkers into which foods to be irradiated are conveyed.

The first commercial facility for irradiating foods in the U.S., Vindicator, Inc., located in Mulberry, Florida, uses cobalt 60 as its source of gamma rays (Lingle, 1992). It is a wet cell irradiator, i.e., the cobalt 60 is housed in and shielded by an 18,000 gal pool of deionized water in a 28 ft-deep well. Palletized food enters the chamber; the irradiation process is activated by raising the cobalt 60 out of the water. When submerged, the source of gamma rays is thoroughly shielded, and workers can enter the chamber safely. Further shielding is provided by 6.5 ft-thick steel-reinforced concrete walls.

Consumer reaction in North America is ambivalent. For example, irradiated Puerto Rican mangoes were enthusiastically received by consumers in Florida, according to press reports (Puzo, 1986; Loaharanu, 1989; Bruhn and Schutz, 1989). A comparison test of irradiated Hawaiian papayas and traditionally processed papayas showed that consumers had no aversion to irradiation (Bruhn and Schutz, 1989). They bought even more irradiated papayas than nonirradiated papayas (Loaharanu, 1989). In January of 1992, irradiated strawberries were sold in the U.S. in Florida (Marcotte, 1992). Despite attempts by protesters to disrupt the market test, consumers did buy the irradiated product and were, in general, favorably impressed by it.

Where then, is the problem for developers? Why isn't there greater availability of irradiated foods? Why, on the basis of the above, do processors believe consumers have an aversion to irradiated foods? Where is their reluctance to irradiate coming from? There is more than a hint of the real problem in the statement in Bruhn and Schutz's paper (1989): "This study [Hawaiian papayas] may not represent actual market conditions because there were no protesters present...." There is deep polarization between proponents of and antagonists of irradiation. This is well documented in the pro/con article authored by Giddings for the pro side and Colby and Savagian for the con side of the irradiation argument (1989). Bruhn and Schutz (1989) suggest that the ultimate acceptance of irradiation will be dependent on the following:

- on whether alternative technologies offer greater or less safety for the food supply,
- on industry's weighing of the costs of failure vs. the potential rewards of success in the introduction of irradiated foods,
- on whether responsible media coverage will accompany the introduction of irradiated foods,
- on whether consumers have been adequately educated to the value of irradiation for them.

To this should be added two more provisos: consumers must see an advantage, a gratification of a need satisfied by irradiation (Best, 1989a), and processors must learn to handle protesters and the "bad press" they can bring with sound crisis management techniques.

Costs for irradiation hover around the $0.05/lb range. Since an irradiation installation is very capital-intense, irradiators will be few and scattered; most of their work will be contracted. Unfortunately, at present, the situation for irradiation could be described as not a race to be first but a race to be second in the market with an irradiated added-value product. It is irradiation trials for test packs that are keeping the facilities for irradiation busy (Lingle, 1992), especially for processed foods at Vindicator, Inc.

Since irradiation doses are restricted to 10 kGy or less, that is, pasteurizing doses, irradiation is most successful in combination with other stabilizing systems, such as refrigeration and CA/MAP systems. In the global sense, irradiation has found very specific market niches (Giddings, 1989). These include dry food ingredients, such as spice seasonings and dried onion and garlic powders; disinfestation of cereal grains, which is by far the largest application, trichinosis control in fresh pork, and the elimination of pathogens such as salmonella microorganisms from poultry. Added to this is the extension of shelf life of soft fruits and tropical fruits.

Irradiation offers great promise for developers of new, minimally processed foods that offer a desired, added-value feature to the consumer.

Some examples of the use of irradiation on added-value convenience products will illustrate some of the advantages and problems that developers must face. Paster and co-workers (1985) used low dose irradiation to pasteurize pomegranate kernels to extend the shelf life of this product and enable it to be used in industrial applications. Complementing this stabilizing system was packaging with nitrogen gas flush and refrigeration. They found no significant difference in taste between irradiated and control pomegranate kernels. However, shelf life was not significantly increased, and a loss of color intensity was noted at all irradiation levels used (2 to 4 kGy). Spoilage microorganisms were greatly reduced, but fungi were found to be more resistant to irradiation. There was one very important observation that should raise a red flag to developers. The dominant fungal contaminant was no longer *Penicillium frequens,* common in nonirradiated kernels; in irradiated kernels, the fungal contaminant was *Sporothrix cyanescens.* Thus, one problem microorganism was removed, only to be substituted by another. *The course of spoilage was altered.* This phenomenon was encountered in CA/MAP techniques. Dempster and colleagues (1985) attempted to extend the shelf life of raw beefburgers, an added-value product. These were packed in vacuum in "Cryovac" bags with irradiation (doses up to 1.54 kGy) and stored at 3°C. Total viable counts of microorganisms were significantly reduced; shelf life was extended by up to 7 days at refrigerator temperatures. However, during storage, irradiated burgers lightened in color and became less red. Both these color characteristics were better than those of the raw, unirradiated control but not as good as the frozen control. On the downside also, irradiation caused an increase in peroxide values in the fat of the burgers and a distinct, unpleasant odor.

Grodner and Hinton (1986) irradiated both sterile and nonsterile crabmeat inoculated with *V. cholerae* with up to 1 kGy. Samples were stored at temperatures ranging from –8 to 4°C for various lengths of time to study the interrelationship of storage duration vs. temperature of storage vs. irradiation dose on the viability of *V. cholerae.* Three interesting findings are discussed in their work, which should demonstrate the complexity of stabilizing systems utilizing irradiation:

1. No *V. cholerae* were found in sterile but inoculated crabmeat nor in nonsterile inoculated crabmeat at dosages of either 0.5 or 1.0 kGy after irradiation. Irradiation at 0.25 kGy could reduce the counts by several log cycles.
2. Pretreatment of the crabmeat (sterilization) was a factor in the survival of *V. cholerae* in unirradiated crabmeat stored at each of the storage temperatures. At the higher storage temperatures (4 and 0°C), the pretreatment removed competing microflora to *V. cholerae*'s advantage.
3. Pretreatment of the crabmeat, however, the authors surmise, caused the loss of some protective factor (perhaps a protein) against freezing,

which was contributory to the survival of *V. cholerae* at the coldest storage temperature. In both irradiated and nonirradiated crabmeat stored at $-8°C$, survival of *V. cholerae* was greater in crabmeat that had not been sterilized prior to inoculation. However, irradiation was effective in reducing the count.

The interrelationships between processing and subsequent stabilizing systems must be appreciated. Developers must be aware that the application of newer techniques of stabilization may alter the traditional course of spoilage. Unless this is understood, developers may find they have replaced the devil they know with the devil they don't know.

H. Hurdle Technology for Preserving Foods

The consumer's awakened interest in the philosophy that fresher is better has led to a proliferation of minimally processed, added-value foods. The description is misleading. These foods can be highly processed (i.e., far from a natural state), employing multiple systems of stabilizing techniques.

By judiciously combining several systems of stabilization selected on the basis of a knowledge of the dominant path(s) of spoilage for some product, technologists can stabilize that product. Each system that is selected produces a minimal change in the food product and is by itself not sufficient to stabilize the food. Each, in combination, complements the stabilizing activities of other systems that are employed. In this way, the high quality shelf life of new (and old) food products can be maximized.

This technique has loosely been referred to as the "hurdle effect" by Leistner and Rödel (1976a), "... multiplication of microorganisms on or in meats does not depend solely on the a_w, but on other factors too, such as temperature, pH, E_h, nitrite and competitive flora. ...shelf life of a meat product is related to the combined effect ('Hürdeneffekt') of several factors...." (See also Leister and Rödel (1976b) for a further general discussion of the hurdle technique with intermediate moisture foods.)

The principle of the hurdle technique has been likened to an obstacle race course set with hurdles of varying heights (different intensities). Some runners (microorganisms) fall at the first hurdle. Those that pass that hurdle are slowed (weakened) for the next one, and more "runners" are felled. Those that survive to face the third hurdle are so weakened that they are in no condition to pass whatever hurdles may remain. The hurdles can be synergistic. That is, two hurdles may complement the preservative action of one another, such that the combined preservative effect of the two systems is greater than the sum of the two systems separately.

This analogy is pictured in Figure 11, where a hypothetical food is shown preserved by four systems. Plant support systems (sanitation, maintenance,

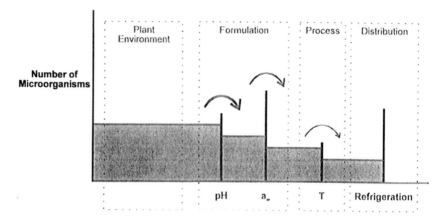

FIGURE 11. Pictorial representation of hurdle technology for a hypothetical food product with four hurdles.

HACCP programs, etc.) keep microorganisms in the plant environment low and consequently keep the product bacterial load low. Two hurdles built in with formulation cause further protection: a lowering of the pH restricts the growth of spores of *Clostridia*; formulation of the product to an a_w of 0.90 provides a further check on microbial growth. Pasteurization (processing) eliminates vegetative microorganisms and stops enzymic activities. Finally, the finished, packaged product is stored, distributed, and retailed at refrigerator temperatures to slow chemical and physical changes.

Grodner and Hinton (1986) extended the shelf life of crabmeat by employing gamma irradiation, modified atmosphere packaging, and storage under refrigerated conditions. In this work, three barriers were used to control spoilage. Even *sous vide* products could be considered as products using a multiple preservative system for their shelf life: partial cooking to provide a measure of pasteurization and blanching action; vacuum packaging to inhibit aerobic microorganisms; refrigeration to check chemical and physical reactions and retard microbial growth. The reader should, however, refer to the comments of Livingston (1990) and Mason and colleagues (1990) regarding the safety of this food service process.

The use of the hurdle technique or multiple stabilizing systems is not new. It has been practiced, advertently or inadvertently, for many hundreds of years and is recognized in many traditional foods, such as:

- sauerkraut (acidification through fermentation, removal of a labile substrate through fermentation, salt addition to control the course of fermentation, presence of highly competitive but benign lactic fermenters),
- pemmican (sun-dried [that is, with lowered water activity] meat to which high acid fruit is added),

- smoked fermented hard sausage (lowered water activity through dehydration, heat treatment by smoking, stabilization by smoke constituents, fermentation that reduces the pH while at the same time removing a labile substrate, preservative action of the lactic fermenters, meat additives such as salts, spices, and herbs, which may have some preservative action),
- cheeses, which are very similar to sausages in their stabilization system (i.e., low water activity, fermentation, perhaps the presence of bacteriocins and competitive flora).

The development of intermediate moisture foods and of the concept of water activity and its relation to food stability sparked the enormous interest that developed for the hurdle concept. Intermediate low moisture foods are foods intermediate in moisture content between dried foods and fresh foods and whose water activity is above that of dehydrated foods but below that of natural foods. They can be consumed without the addition of water.

Neaves and colleagues (1982) at the then British Food Manufacturing Industries Research Association (now Leatherhead Food Research Association) studied preservative interactions to inhibit the growth of *Clostridium botulinum* in a pork medium. The stabilizing systems they used were preservatives and preservative combinations (potassium sorbate, sodium benzoate, or ethyl *p*-hydroxy benzoate), humectants and humectant combinations (salt or propylene glycol), pH, incubation temperature, and pasteurization.

It is important for new product developers to note Neaves and co-workers' observation that, although salt plus sorbate combinations were effective in inhibiting growth and toxin production, a lowered pH or reduced temperature was necessary to complement the preservative/humectant stress. In their opinion, inadequate preservation is far more dangerous than no preservation in a food system, since the former could result in the unwitting ingestion of toxic food; they found that toxin could be present in a food when there was no obvious appearance of spoilage.

Webster and co-workers (1985) used a different model system on which to test 72 stabilizing variable interactions against what they termed a "challenge cocktail" of microorganisms. This cocktail consisted of *Staphylococcus aureus, Bacillus subtilis, Pseudomonas aeruginosa, Streptococcus faecalis, Lactobacillus casei* var. *rhamnosus,* and *Clostridium perfringens,* all well-recognized food spoilage or food intoxicating microorganisms. The variables were four different levels of a_w, four different pH values (5.5 to 7.0), plus the presence of sodium citrate and sodium benzoate.

The important point for product developers to remember is Webster and co-workers' observation that when a lowered a_w was the sole hurdle, the predominant microorganism was *Staphylococcus aureus.* When the added stress, i.e. hurdle, of citrate/benzoate was added to that of reduced a_w, the dominant microorganism became *Streptococcus faecalis.* Thus, the hurdles that are chosen can greatly alter the course of microbial growth. Developers must understand the stabilizing system in its entirety.

Carlin et al. (1990) demonstrated very neatly how the course of spoilage could alter with the hurdles that were imposed. They studied the influence of controlled atmospheres (CO_2, 0.03 to 40%: O_2, 21 to 1%) and storage at 10°C on the spoilage of commercially prepared, fresh, ready-to-use grated carrots available on the French market. This temperature is higher than that recommended by the French government for such products. Nevertheless, it was the temperature at which ready-to-use products were frequently sold. The growth of both lactic acid bacteria and yeasts was more rapid as the CO_2 concentration increased from 10 to 20%, regardless of the oxygen concentration present. The composition of the atmosphere in CA/MAP products can influence the population dynamics of spoilage.

The high levels for some humectants required to lower a food's water activity preclude their use in many food products for flavor considerations. There are, nevertheless, a number of humectants, including proteins, protein hydrolysates, amino acids, and sugars and sugar alcohols, that can be used to depress the food's water activity (see, for example, Table 2, in Chirife and Favetto, 1992).

Chirife and Favetto (1992) also remark on the observation made by others that, in addition to the a_w lowering of various solutes/humectants, there may be very specific solute/humectant effects. For example, ethanol has an antimicrobial effect that is not due solely to its a_w-lowering ability. Sodium chloride and sucrose inhibit the growth of *Staphylococcus aureus* around a value of 0.86. When other solutes, such as diols and polyols, are used, growth is inhibited at a much higher a_w. Therefore, developers must be aware not only of the hurdle imposed by a lowered water activity but must also recognize the influence of the humectant/solute system that produced that level.

There are precautions that must be taken in the application of hurdle technology. It requires that high levels of quality control, sanitation, and hygiene be exercised in the processing plant. GMP and HACCP programs must be exercised scrupulously. The hurdles established for any product are not designed to protect against high microbial loads: they could be swamped if the plant environment during processing and handling prior to packaging contaminated the product. Leistner (1985, 1986, and 1992) and Leistner and colleagues (1981) emphasize that the technology of the hurdle effect can be influenced by high microbial loads.

Developers must, further, fully understand the properties of the food product under development and understand the possible destabilization mechanisms of that food product with respect to intoxication and spoilage. The application of hurdle technology is, in effect, a hazard analysis of the product, in conjunction with the application of critical controls built into the product.

The final precaution when using this technique is that the developer must understand (and record in the product description) the purpose and function of each and every ingredient or additive in the product. This alerts future developers faced with reformulation of the product to the importance of each and every ingredient. The stabilizing system is a balance of stresses that is unique for that product composed of those particular ingredients and processed precisely in the manner that it was.

Hurdle technology, use of multiple preservative systems, offers the opportunity to prepare novel, minimally processed, added-value food products, which can be shelf-stable at room temperature. Usually, however, refrigerated handling and storage are additional hurdles used in conjunction with selected combinations of heat treatment, reduced a_w, reduced pH, preservatives, antioxidants, MA/CAP or interactive packaging, reduced E_h, or irradiation.

I. Low Temperature Stabilization

Chilled, prepared foods, also referred to as REPFEDs, refrigerated processed foods of extended durability [see, for example, D. A. A. Mossel cited in Scott (1987) and Brackett (1992)], were mentioned with the topic of hurdle technology. They are particularly sensitive to temperature abuse and are susceptible to hosting pathogenic psychrotrophs. Consequently, they are usually preserved using hurdle technology (Scott, 1987). Their growing acceptance by consumers justifies a separate discussion.

The shelf life of REPFEDs is, as one would suspect, highly variable. It depends on the intrinsic properties of the food, the properties of the ingredients composing it, the stabilizing systems used in its design, and extrinsic factors.

Bristol (1990) reported that many of these products on the market had anywhere from 12 to 60 days of shelf life, with most in the 20- to 30-day range. Fresh pasta at 40 days and the accompanying sauces at 60 days were the most stable. However, the biggest problem with chilled added-value foods arises because of the lack of control in the distribution and handling systems. This includes both the transport system and conditions at the retail level. Clarke (1990) recorded the day-to-day temperature variations in multideck display units in retail outlets. During defrost peaks, temperatures as high as 15.8°C were recorded, and these could go as low as −1.2°C; yet mean temperatures were satisfactory.

Lechowich (1988) reported that low-acid foods (pH > 4.6), with mild heat treatment and vacuum packaging or CA/MAP, had approximately 14 days of shelf life. Acidic foods were less of a problem. *Sous vide* products had, in general, 2 to 3 weeks of shelf life when held at 2 to 4°C.

The short shelf life of chilled foods is one factor against their wider acceptance. Both Lechowich (1988) and Day (1989) in the U.K., where chilled foods are much more widely accepted, cited the inadequacy of vacuum packaging or MA/CAP plus refrigerated storage and distribution alone to stabilize chilled foods. Both recommended that these stabilizing systems be supplemented with acidification, use of competitive flora, addition of preservatives, or heat processing.

Developers would do well to heed Waite-Wright's (1990) recommendation for the need for a five point program to ensure high quality chilled products that are safe to eat:

- good raw materials, the procurement of which requires close communication with reliable and dependable suppliers, plus audits of all supplies,
- a safe controlled process employing all the support systems of the plant,
- strict vigilance against cross-contamination caused by workers, or poor separation of cooked and raw materials and their associated equipment,
- close adherence to a sound plant hygiene and sanitation program,
- employee training in sanitation and personal hygiene.

All are necessary for safe chilled foods.

Refrigeration as a technique to extend the good shelf life of foods is coming under much closer scrutiny. Concern for the microbiological stability and safety of chilled foods has grown in the last few years. The realization that a number of pathogenic microorganisms and spoilage microorganisms, which pose a potent hazard in chilled foods, can grow and survive at so-called good refrigerator temperatures has sparked this concern. Developers of chilled food products must control potentially harmful microorganisms in the design and formulation of their products and must establish that handling, processing, storage, and distribution controls are in place to prevent recontamination.

J. Summary

In summary, the foregoing review of developments in stabilizing food products has demonstrated many opportunities for producing new products. Food technologists must be very circumspect in choosing which of these techniques to use and how they are combined in any particular product. The danger lies in two areas. First, rates of spoilage, whether they be of physical, chemical, biological, or microbiological origin, do not proceed uniformly under any given set of stresses to which food products may be subjected. This is depicted in Figure 12. Rates of loss of some quality attribute, of growth of pathogens, and of spoilage by yeasts and molds are shown plotted against acceptable shelf life of a hypothetical product. Below 10°C, obvious microbial deterioration alerts consumers to a problem. But when the storage temperature is higher, pathogens may grow and be present in sufficient numbers to be a hazard. Neither the quality attribute of the product nor microbial spoilage forewarns of a problem.

The second danger for developers is this. The course, or mechanisms, of product destabilization may be altered in the application of the newer technologies or the combinations of them that may be applied to foods. A technique applied to a food product with a known microbial spoilage pattern may allow some hitherto unsuspected microorganism to emerge as the dominant spoilage

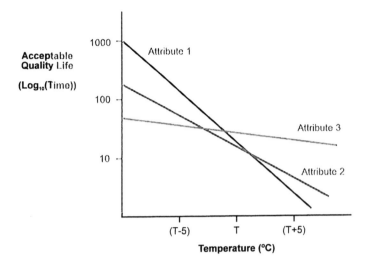

FIGURE 12. Deteriorative rates of various quality parameters of a hypothetical product with respect to temperature.

cause. Or a technique (e.g., ohmic heating) to pasteurize a product may cause physical changes in a product that are esthetically unacceptable to consumers. Great care must be used in selecting stabilizing systems for new products.

IV. MEETING THE CHALLENGE: NEW INGREDIENTS

For the purposes of this text, no clear distinction will be made between an ingredient and an additive. Readers should be cautioned, however, that in the legislation of many countries, such a distinction is made. It is unfortunate that not always do these distinctions correspond from country to country. A substance may be classed as a natural substance in one country and hence as an ingredient but be classed as an additive in another. Similarly, the concept of a natural product and a nature-identical product can be widely divergent from country to country. For the purposes of this text, anything deliberately added to a food will be considered, rather simplistically, as an ingredient.

A word of caution, therefore, is in order when ingredients are to be used in any new food. Regulations respecting the use of ingredients vary from country to country: acceptance in one country does not guarantee that the ingredient will be accepted in another. Therefore, mention of ingredients in this text does not imply that they are permitted but merely illustrates their potential applications as ingredients. Whether a product under development is meant for export markets or not, regulations respecting the status of all ingredients to be considered in its formulation should be clarified. Even when the ingredient is permitted, there can be restrictions regarding usage levels and types of foods in which the ingredient can be used. Many food products have standards of identity — in the U.S., more than 300 foods do — which may limit substitutions of ingredients.

The use of some ingredients may have an impact on the marketing department's promotional plans. This must be discussed early in development with marketing personnel. For example, to make "lite", low- or no-fat, high-fiber, or low-calorie variants of foods that possess a standard of identity may cause conflicts in labeling. If export markets are goals for the marketing department, consideration must be given to the fact that labeling laws respecting nutritional claims are different in many countries (for example, between Canada and the U.S.). The consumers' interest in "green" labels (their desire for natural ingredients) and the marketing personnel's obvious desire to please consumers will limit the ingredients that technologists can choose in formulations. Development teams need to consider carefully all the implications for labeling, for promotional claims, for export markets, and for environmentally concerned consumers that may be entailed in the use of ingredients. This should be assessed early in the development process to prevent useless work later.

Walking through the aisles at many food trade shows could easily lead one to think that ingredient suppliers believe that consumers are more interested in what is not in their foods than in what is there. Exhibitors demonstrate the use of ingredients that are low-calorie, fat-free or low-fat, sugar-free, low-sodium, cholesterol-free, and without saturated vegetable fats. On the positive side, fiber, soluble or insoluble and derived from every conceivable source, is "in", as are fat-replacers or fat-mimetics, many of which are derived from fiber sources.

All these ingredients give food technologists opportunities to develop a wide variety of nutritional or dietetic alternatives to existing traditional products.

A. Fat Ingredients

The nutritional guidelines of many countries recommend that their citizens reduce the amount of calories derived from fat. Thus, today's consumers are interested in reduced calorie foods, due to their government's endorsement, their own awareness that obesity is related to many disease conditions, and the strong suggestion reported in the media that high fat diets and breast cancer may be directly associated. Weight for weight, fat contributes more than twice (9 kcal/gm) as many calories as protein or carbohydrate (both with 4 kcal/gm). It would appear that the easiest route to lower calories would be to eliminate or lower the fat content of food.

The developer has few techniques with which to lower the fat content of foods:

1. reduce or remove the fat by trimming, skimming, rendering, or extracting the food,
2. reduce or replace the fat in the food and substitute for the fat removed, wholly or in part, some less calorically dense material (fat extender). Fat extenders may still provide calories, but weight for weight they provide fewer calories than the fat they replace,
3. reduce or replace the fat content with some nonabsorbable substance (fat mimetic or replacement) with fat-like properties.

The first solution is impractical. The resultant food product would be greatly altered by the physical treatment. It might even, perhaps, become unpalatable, since fat provides many organoleptic properties to food. Solutions two and three require a substance, a fat-extender or mimetic, that provides all the sensory properties of fat and is nontoxic. It is in this area that developers of new food ingredients have concentrated their efforts.

Fat replacements or mimetics have been derived from a number of products. They can be very roughly grouped into three classifications:

- replacements based on physically or chemically modifying some polysaccharide fraction such as cellulose or dextrins. These use as a starting material such carbohydrate material as, for example, rice bran, pea fiber, oat fiber, corn starch, or tapioca dextrin,
- replacements based on altering the physical character of some natural protein [for example, Simplesse™, a trademark of The NutraSweet Company, which was originally made from milk solids and then milk and egg solids, Bertin, (1991)] by creating a suspension of protein microparticles, which has the mouthfeel characteristics of an oil,
- replacements synthesized, some from natural foods such as sucrose, or other sugars esterified with fatty acids, which have fat-like properties. Silicones and long chain members of alkane hydrocarbons with characteristics resembling fats have been used as fat substitutes (LaBarge, 1988) and can be included in this category.

The synthetic or *engineered* fats, as Singhal and co-workers (1991) referred to them, owe their classification to the way in which the basic skeleton of a fat molecule is modified. Thus synthetic fats and oils could be modified by:

- replacing the tri-hydroxy glycerol fraction with other polyhydroxy compounds. Sucrose, raffinose, and amylose have been esterified with natural fatty acids. Propylene oxide can serve as the backbone. Sucrose polyester (Toma et al., 1988), one such fat replacer, is nonabsorbable and also able to reduce the cholesterol level in the body. Other replacements provide fewer calories per gram than do fats and qualify more as extenders,

- substituting other acids, such as branched carboxylic acids (sterically hindered fatty acids), dicarboxylic acids (fumaric, succinic) or shorter-chained fatty acids for some or all of the long-chain fatty acids of the triglyceride structure. Medium-chain triglycerides are an example of this category. They are specialty fats prepared by the hydrolysis of vegetable oils, the subsequent fractionation of the fatty acids, and re-esterification of these fatty acids to a glycerol backbone (Megremis, 1991),
- substituting a polycarboxylic acid for the glycerol backbone and then esterifying suitable long chain alcohols to the acid. (This has been referred to as reversing the ester linkage.) Common acids used for the backbone are citric acid or tricarballylic acid, the tricarboxy acid of glycerol. LaBarge (1988) discusses these and feeding trials using them in his review,
- reducing the ester linkages of the triglyceride to ether linkages and thus changing the properties of the fat. LaBarge (1988) describes these ingredients as much more slowly susceptible to hydrolysis than the esters but more prone to oxidation.

(Food product developers are cautioned that mention of these ingredients should not be interpreted as meaning that all, or any, countries have necessarily sanctioned the use of these products. Developers must verify whether the ingredients they use are permitted in the class of foods they want to use them in and in the amounts they wish to use.)

Medium-chain triglycerides are worthy of further note. These may have limited value as fat replacements where only caloric reduction is the major requirement. However, as Megremis (1991) reports, they also serve as processing aids by acting as flavor carriers and are used in confections where their low viscosity can prevent sticking and provide gloss. Their biggest value comes in their usefulness in specialty nutrition products, a potential niche market. Medium-chain triglycerides are absorbed by the portal system because of their chain length. Consequently, they are metabolized in the liver as rapidly as is glucose (LaBarge, 1988). As a result, these triglycerides are desirable in special diets to provide a rapid and concentrated source of energy to people with intestinal malabsorption problems (Crohn's disease, colitis). This facet of their usefulness is described by Babayan and Rosenau (1991). They further describe the use of a medium chain triglyceride oil (chiefly caprylic and capric acids for the triglycerides) in cheddar-type and fontinella-type cheeses.

While noting the many nutritional advantages of medium-chain triglycerides for patients with malabsorption diseases, Kennedy (1991) points out that some long-chain fatty acid triglycerides are needed to meet the essential fatty acid requirements of the body. Therefore, structured lipids (interesterified lipids formed from medium-chain triglycerides and long-chain triglycerides) provide unique benefits, which Kennedy (1991) summarizes, in part, as:

- improving immune function,
- lessening cancer risk,
- lowering cholesterol.

Jojoba oil, a natural oil from the seeds of a hardy bush of the southwestern U.S., has also been explored as a possible substitute for fat (Hamm, 1984). Its oil is a liquid wax comprised of esters of fatty acids and alcohols (Anon., 1975). It does solidify at refrigerator temperatures (Hamm, 1984), and this may limit its use in some foods.

Gums also serve as aids in fat-reduced foods by stabilizing emulsions and providing viscosity, thus simulating some properties of oils and permitting the reduction of oil in the formulation. Dziezak (1991) describes the properties of gums and reviews their applications in emulsification.

There are problems with the use of fat replacers/extenders in foods. Product developers should carefully assess the properties of each before using any in products. Some, like the alkoxy citrates, are not thermally stable (Singhal et al., 1991), some may reduce the absorption of other macro- and micronutrients (LaBarge, 1988), and some may result in anal leakage when consumed (LaBarge, 1988). The major problem presented to food technologists is to determine which substitute will be stable and most suitable in the particular food system in which it is to replace fat.

Dziezak (1989) provides a very readable capsule review of some of the properties of natural fats and oils for technologists unfamiliar with fat chemistry. Her accompanying treatment on fat substitutes is now outdated. However, as she points out, this area of ingredient development is undergoing tremendous change.

Product developers have rapidly reformulated products containing saturated tropical oils to remove these ingredients from products. This occurred when the prevailing wisdom was that these saturated vegetable fats behaved as saturated animal fats in the body. Berger (1989) describes concisely the situation concerning the use of tropical oils in the U.S., presents a cogent argument for their continued use, and disputes attacks against their bad nutritional value.

B. Sugars, Sweeteners, and Other Carbohydrate Ingredients

Several years ago, sucrose, from sugar cane or sugar beets, was the main sugar used in food processing. Certainly, other sugars have been used historically: honey and maple syrup have been common for centuries. Sherman (1916) in his classic, *Food Products* (1916 printing), describes only the manufacture of cane sugar, beet sugar, molasses, refiner's syrup, maple syrup, open kettle cane syrup, and honey. Wiley (1917) included sorghum syrup. In the past, these were the sugars in common use. Nevertheless, sucrose was, by far, the main sugar in food processing.

There were many reasons for the popularity of sucrose in manufacturing. It was readily available and, therefore, comparatively cheap. It served many functions in foods in addition merely to sweetening them. (These will be explained below.) Along with the advantages, the use of cane sugar also brought disadvantages. Sucrose is cariogenic. Health problems associated with high caloric intake from refined sugars, such as sucrose, typically found in North American diets have become more apparent to consumers. But these are later developments. However, these disadvantages have led to the search for substitutes.

Alternative sweeteners to reduce or eliminate the need for sucrose in food products fall into two categories:

- those that are intense sweeteners in their own right but contribute no or very few calories to the food. These can totally replace a sugar *where sweetness is the only function demanded of the sugar*;
- those sweeteners which are caloric alternatives, i.e. on a weight basis they are more intense sweeteners than is sucrose but also contribute other properties, including calories, associated with sugars. These are frequently referred to as the bulk sweeteners (Giese, 1993).

The low-calorie group includes:

cyclamates	saccharin
sucralose	aspartame and other peptides
glycyrrhizin	miraculin
monellin	thaumatins
stevioside	rebaudiosides
neohesperidin	dihydrochalcones
acesulfame K	L-sugars

As with other categories of ingredients, caution must be taken in choosing intense sweeteners, since some alternative sweeteners have limited stability in certain food systems; and some produce a slow onset of sweet sensation, while others have bitter or undesirable aftertastes. Care is required to select the sweetener most suited to the food system where it will be employed.

The other alternative group comprises crystalline fructose, high fructose corn syrup, mannitol, sorbitol, xylitol, hydrogenated sugars (e.g., maltitol), hydrogenated starch hydrolysates, and isomalt. This group is, by far, the more important for developers. Members of this group possess many of the unique functional properties of sugar in addition to sweetening.

Sugars provide more than sweetness. They provide mouthfeel (viscosity) in some beverages. They are used to regulate a food's water activity. They influence texture in foods. They can stabilize color. Sugars serve as bulking agents in hard candies and other confections; they provide unique humectant properties in jams and jellies and baked goods; they assist, along with pectins, in the control of gel texture for jams, jellies, and marmalades; and they can be used to control graining in soft-centered candies.

The uniqueness of sugars, in particular sucrose, is best exemplified by a beautiful demonstration — beautiful because of its simplicity — described by Chinachoti (1993). In what she called her "magic trick", she mixed corn starch and water. Mixing produced a wet powder, an unpourable wet powder. Next she added another dry powder to the wet powder. Mixing produced not an even drier mixture but eventually a flowable slurry. The second powder was ground crystalline sucrose.

The corn starch/water mixture, despite its dry appearance, has a high a_w (approximately 0.98). With the addition of the sucrose, the dry appearance disappears after mixing. The seeming solid becomes a liquid with an a_w that has dropped to 0.93. As magical as this may appear, Chinachoti (1993) went a step further. One would expect the corn starch/water mixture to spoil faster microbiologically than the corn starch/water/sucrose mixture if one's prediction is based solely on the two water activities. Chinachoti demonstrated that this was not so. The mixture with the lower a_w spoiled more rapidly. The addition of sucrose had indeed lowered the a_w, but she demonstrated that it had also increased the amount of mobile water. As Chinachoti explained, "... sucrose undergoes a phase change from crystalline to dissolved sucrose upon hydration ... solvation of sucrose promotes mobility and helps facilitate the spoilage."

The lesson for product developers is this: sucrose and other crystalloids contribute to a lowered water activity, but it is simply not adequate to measure water availability to predict product stability. Other factors, such as water mobility and phase transitions, may play a more important role than does water activity alone in product stability.

The importance of phase transitions and the physical state of food components, in particular sugars, is treated in depth by Roos and Karel (1991a, b), Noel and co-workers (1990), and is reviewed by Best (1992). Noel et al. (1990) discuss the importance of phase transitions and the glassy state in the storage of frozen foods, in freeze dehydration, and as a cryoprotectant for foods. If the aqueous system of a food can be so formulated as to maintain the matrix in the glassy state, then water crystallization and damage caused by crystal growth can be minimized.

For example, the drying of pasta products can be greatly improved by close control of humidity, temperature, and drying rate to control glass transition. This maintains pasta in the glassy state. The quality of extrusion-puffed snack products owes much to control of phase transitions. Roos and Karel (1991a, b) suggest that glass transition temperatures may be a factor in browning. Grittiness of lactose in ice cream, caking in milk powders, and stickiness of hard candies are all related to the behavior of ingredients in the glassy state (Noel et al., 1990). Best (1992) provides an excellent overview article reviewing the as yet far from established understanding of water's role in product development. (See also Slade and Levine, 1991, for a very extensive discussion on phase transitions in food and their role in stabilization.)

C. Fiber Ingredients

Heightened awareness of the importance of fiber in the diet dates back to the mid 1980s, when the National Cancer Institute (U.S.) promoted the potential cancer prevention benefits of a high fiber diet. This campaign alone gave a boost of an estimated 30% in sales to one well-known ready-to-eat bran breakfast cereal (Anon., *Prep. Foods,* 1988). However, the beneficial use of high fiber diets in aiding regularity has been promoted for many years, and its laxative effects have been recognized for many more years. Rusoff (1984) reviews the physiological and nutritional effects of dietary fiber in man. Ripsin and Keenan (1992) discuss more recent data from nutritional studies specifically centering on oat products and their value in reducing blood cholesterol levels. Wood (1991) reviews the physicochemical properties of oat betaglucan. For developers wishing to obtain some background in this field, Schneeman (1987) reviews the physiological responses of the various types of fiber when ingested (soluble vs. insoluble fiber).

Health claims have been made that state that some fiber, in particular certain sources of fiber, such as oat, rice (Normand et al., 1987), pectin (Reiser, 1987), and guar, can clear cholesterol from the circulatory system. This implies that there is a role for fiber in lowering the risk of cardiovascular disease. Furthermore, it is speculated that fiber may play an as-yet-undetermined role in the prevention of some cancers. For these reasons, products with fiber have significant consumer appeal. Demand for products with high fiber content has grown as food companies try to satisfy this increasing awareness by consumers that food and health go together. Fiber is now emphasized in a wide variety of foods, from baked goods such as breads, muffins, and cookies, to pastas, beverages, and processed meat products.

Dietary fiber is an all-encompassing term for a group of natural plant components, which includes the carbohydrates, cellulose, hemicelluloses, pectins, gums, and a noncarbohydrate component lignin; none of these are themselves compounds of fixed composition. Dietary fiber (roughage, bran, unavailable carbohydrate) is a poorly defined entity, which in addition to the above, may, in its crude form, have other plant materials with it. This lack of uniformity is a disadvantage because developers find that it results in variable properties in the products they create. This apparent disadvantage can be overcome by purification and blending to produce the desired uniform properties. Cellulose is a linear polymer of varying lengths, consisting of glucose units. Hemicelluloses are a complex mixture of linear and branched polysaccharides with side chains composed (most often) of xylose, galactose, and other hexose and pentose sugars. Pectins are equally complex mixtures of polygalacturonic acid units, as are the plant gums and mucilages. For example, the seed gum of mesquite has been determined to be a galactomannan (Figueiredo, 1990). Lignin is a polymer of units of phenyl-propane.

Dietary fibers come from many different readily available plant sources. Some arise as byproducts of processing the raw plant. Some sources of fiber are prunes, corn, oranges (Hannigan, 1982), soybeans, peas, oats, rice, spent barley, sugar beets, tofu processing, various vegetable gums, such as gum arabic, guar gum, locust bean gum, and mesquite (Figueiredo, 1990), wood, potato peelings, carrageenan, and psyllium grain husks (Anon., 1990b). Chitin from the shells of crabs or from microbial sources and chitosan prepared by the deacetylation of chitin, are biopolymers with promise as sources of fiber. The preparation and properties of both are described by Knorr (1991), with possible applications in food processing and waste management.

Each source produces fibers with unique properties, and when used in combinations, many interesting applications in food systems are possible. There are many technical considerations in addition to any nutritional benefits to be pondered when choosing a fiber or combination of fibers to serve some functional purpose in a food. Fiber, as Penny (1992) points out, will affect color, flavor, oil and water retention (hence the a_w of foods), rheological properties such as colloidal/emulsion stability, texture, gel-forming properties, and other viscometric properties, such as thickening and mouth feel of products. They can be used to control crystal formation in sugars or to act as cryoprotectants in freeze/thaw food systems. For example, Ang and Miller (1991) list some of the many products in which one fiber, cellulose powder, is used:

bread	pasta
cookies	cakes
rusks	cheese
soups and sauces	yellow fat spreads
comminuted meats	meat and fish pastes
slimming foods	dietetic products

They also point out the growing interest in cellulose powder in stabilizing frozen food analogues, such as surimi. Both Penny (1992) and Ang and Miller (1991) detail specific applications of fiber in foods.

Characteristics to be aware of in the selection of fiber for use in a food application are, of course, dependent on what function the fiber is to serve in the food. The two most important characteristics are particle size for texture considerations and soluble-to-insoluble fiber content ratio. [See Olson and co-workers (1987) for a discussion on analytical procedures for soluble dietary fiber and information on total dietary fiber and insoluble dietary fiber.] Soluble fiber has excellent water-binding properties and as such will have a pronounced influence on the texture (moistness) of baked goods in particular, on the stability of products, and on the viscosity in beverages. In general, soluble fiber is highest in vegetable gums and lowest in cellulosic fractions of plant materials. From the natural sources available, high to low ratios of soluble to insoluble fiber may be obtained to suit whatever purposes developers have in mind.

A particular fiber can frequently be used synergistically with other fibers or with starches. Arum root, from which konnyaku, a vegetable jelly (Downer, 1986) that for many centuries has been a mainstay of Japanese cuisine, is prepared, is the source of konjac flour, a glucomannan. Tye (1991) reviews the synergism of konjac flour with kappa carrageenan and starches in maintaining the structural integrity of shaped foods during thermal processing and to simulate the textural properties of fat and connective tissue in sausage-like products.

Crude fiber should not be equated with dietary fiber. Crude fiber consists largely of cellulose and lignin remaining after the treatment of a food material with sulfuric acid, sodium hydroxide, water, alcohol, and ether.

D. Textured Proteins

Vegetable proteins, textured to simulate the fibrous structure of meat, have been available for several years. They were popular when meat prices soared in the U.S. during the latter part of the 1970s and textured vegetable proteins were used to extend meat products. When meat prices dropped as the beef supply grew, many of these products failed, partly because they did not meet consumers' expectation and partly because consumer demand proved to be a fad. (This underlines a point made earlier. When new products are being developed, market evaluation must determine whether trends are fads, that is, momentary aberrations from normal behavior, or whether they are true changes in behavior on the part of the consumer.) Textured proteins are still widely used in many products, especially dried soups.

The growing green movement is associated with many trends, not the least of which is a growing vegetarian sector of the population. Textured vegetable proteins have had a reasonable success with main dish items for vegetarians. A mycoprotein, developed by Rank Hovis McDougall, has been successfully texturized and used by the Sainsbury company (J. Sainsbury plc, U.K.) in a series of meat-flavored products. In test market locations, these have been well received (Godfrey, 1988). These products have a dual advantage over meat, which marketing personnel can promote: they contain no animal fat and they contain fiber. The mycoprotein is currently being tested by Sainsbury for vegetarian (i.e., non-look-alike meat) dishes.

Another development in texturing of fish protein is hardly new. Surimi is freshwater-leached fish muscle. Fish muscle is washed repeatedly to remove soluble proteins and fat. Lee (1984) has described its processing in detail and its application in some products. Lanier (1986) discusses the functional properties of surimi, particularly its excellent gelling properties. Surimi has been used successfully with crab meat, shrimp meat, and smoked salmon in fabricating crab legs, butterfly shrimp, and lox, respectively. Sausages, made with surimi, have been very popular in Japan but have not been as successfully received in North America.

These products enable developers to create a wide range of added-value products with unique textures and forms.

E. Antioxidants

Ingredients (and the ingredient list that results) have become concerns of consumers. In turn, they are a concern for companies who wish to promote their products with as "green" an image as possible (i.e., as natural or made from natural ingredients). In a survey of food R & D executives, *Prepared Foods* reported that a significant percentage of executives "... expected their companies to increase efforts to reformulate ingredients with "additive" label connotations out of existing product lines" (Anon., 1990a). Stabilizers, preservatives, emulsifiers, texturizers, and the many other ingredients that must go into a food product to provide a high quality shelf life or to attract and whet the appetite are very necessary in product formulations. If only they were all natural!

Many herbs, spices, and foods contain natural preservatives that can extend the shelf life of foods. This has been known for many years, both in the art of cooking and later in the science of preserving food. Scientists found that some food components that occur naturally have antimicrobial and antioxidant properties. Thus these "natural additives" in foods gave rise to the hope that they could be used to give foods an extended shelf life. They can, but as might be expected, the very strong flavoring that herbs and spices, in particular, contribute to a food, do limit their use.

Kläui (1973) lists a number of herbs, spices, and other foods, as well as nonfoods, which have antioxidant properties:

oregano	rosemary
nutmeg	mace
sage	turmeric
allspice	cloves
marjoram	thyme
tomatoes	carrots
citrus fruits	buckwheat
tea leaves	onions
beer malt	heated gelatin and casein
coffee beans	wine

Extracts of alfalfa, oats, rice bran oil, rice germ oil, and pea flour were also reported to display antioxidant properties (Kläui, 1973). However, the efficacies of the antioxidant properties of the various natural materials seem to depend on the fat system in which a particular material is tested.

Davies et al. (1979), from the Leatherhead Food Research Association, added cocoa butter, cocoa powder, cocoa shell, defatted cocoa powder, and oatmeal to the list of natural materials with antioxidant properties. They tested these materials, as well as rosemary and sage, and various extracts of them, against the oxidative stability of corn oil, coconut oil, a blend of corn oil and coconut oil, olive oil, cottonseed oil, tallow, and freshly rendered lard. They demonstrated that all have antioxidant activity, but again, "… the antioxidant activity of a particular material and its extract appears to depend upon the nature of the oil or fat used for testing and the type of solvent used for extraction of the antioxidant components."

Pokorný (1991) reviews the natural antioxidants under the following topics:

tocopherols	from sesame oil
phospholipids	from olive oil
phenolic compounds	from cereals
flavonoids	from herbs, spices, and algae
carotenoids	polysubstituted organic acids
proteins, peptides, amino acids, and Maillard products	

These topics provide some idea of the wide variety of natural substances with antioxidant activity. He concludes his review, however, by suggesting that the best approach for protecting foods against oxidation is to avoid the use of either natural or synthetic antioxidants. This could be accomplished by removal of oxygen, by avoiding or eliminating the sensitive substrate, or by decreasing the oxidation rate by low temperature storage and avoiding light.

An undated report from the Specialty Food Division of Ingredient Technology Corporation (Woodbridge, NJ) presents evidence of natural antioxidants in molasses and attempts to characterize these.

Individual components of some natural ingredients with antioxidant properties are well known. They have been incorporated into some commercial antioxidants. For others, the active antioxidant agent is as yet unknown. Because of this, many companies are vying with one another to come on the market with new ones. Kalsec, Inc. (Kalamazoo, MI) has developed a commercial antioxidant extract from rosemary, which has found good use in meat products (Anon., 1990a). The active component in cloves is eugenol. Unfortunately, eugenol's strong clove odor limits its use to baking products. Its isomer, iso-eugenol, has a more acceptable odor — woody and spicy (Heath, 1978). This is also found naturally in nutmeg, basil, and other oils, according to Kläui (1973). Other active compounds contributing to antioxidant properties are tocopherols and carotenes contained in spices and herbs. Gordon (1989) has also reviewed the natural antioxidants and attempted a brief outline of some of the mechanisms involved, which can assist developers' understanding of their use.

F. Antimicrobial Agents

Some spices and herbs, as well as other natural materials, have been known for some time to have antimicrobial activity against a wide spectrum of genera of microorganisms. Some microorganisms are surprisingly sensitive to some natural components. Beuchat (1976) found that *Vibrio parahaemolyticus* was very sensitive to oregano. Shelef et al. (1980) screened over 40 microorganisms, which were either Gram-positive or Gram-negative genera associated with foods, for sensitivity to sage, rosemary, and allspice. Allspice was the least effective as an antimicrobial agent, while both rosemary and sage demonstrated antimicrobial activity. Sage proved more stable than rosemary in its antimicrobial activity over time. Generally, the spices were more effective against Gram-positive microorganisms. The authors (1980) ascribed the antimicrobial action to cyclic terpenes present in the spices.

Zaika and Kissinger (1981) studied the behavior of oregano against *Lactobacillus plantarum* and *Pediococcus cerevisiae* alone and in mixed culture. They noted both inhibitory and stimulatory effects with the spices. At low concentrations of oregano, there was stimulation for both microorganisms. As the concentration rose, an inhibition of both acid production and viability was observed. The authors were able to separate the inhibitory and stimulatory factors.

Zaika and co-workers (1983) extended their work to a study of oregano, rosemary, sage, and thyme against lactic acid bacteria. Again, microbial growth and acid production were retarded as concentrations of these herbs were increased. They did observe one point of interest to serve as a caution for developers. Resistance to a particular spice could be induced in microorganisms. When this occurred, the microorganism in which resistance was induced was also resistant to the other herbs. This suggests that the inhibiting agents are identical in these herbs and spices.

Beuchat and Golden (1989) reviewed antimicrobials in foods under the following headings:

enzymes and other proteins	organic acids
medium chain fatty acids	plant essential oils
pigments and related compounds	
humulones and lupulones	oleuropein
hydroxycinnamic acid derivatives	
caffeine	phytoalexins
theophylline and theobromine	

As with the antioxidants, these headings point to the wide diversity of compounds and sources with antimicrobial activity. The phytoalexins, also reviewed briefly by Mertens and Knorr (1992), are receiving a great amount of interest. They are substances produced by many plants to protect themselves when they are stressed by microbial infections, injury, or other traumas.

Sage, rosemary, allspice, oregano, thyme, and nutmeg seem to be the herbs and spices of greatest applicability as antimicrobial agents for foods. Oregano and thyme both contain oils high in thymol and carvacrol. Sage and rosemary can be grouped together as herbs containing cineole in large concentrations. Allspice is high in eugenol (Heath, 1978). All these components in these spices and herbs are substituted phenols, which may explain the antimicrobial activity.

Food technologists may be able to take advantage of the preservative action of some natural ingredients to prolong the acceptable shelf life of some new products. Although the use of spices and herbs is limited by their strong flavors, the trend to low-salt foods, the growing popularity of ethnic foods, which can be highly spiced, and the popularity of vegetarian foods, which because of their blandness also are usually spiced, are making highly flavored foods more acceptable. The high flavor impact from the use of spices and herbs with other potent flavorants such as vinegar, lemon, onions, shallots, and garlic can mask the lack of salt and the blandness of many products.

A class of antimicrobial substances that are receiving much interest for their possible use as natural food preservatives is bacteriocins. These are nitrogen-containing substances with potent antimicrobial activity. Fermentation, and particularly lactic fermentation, has been well established historically as a stabilizing process, as has the use of acids such as lactic or acetic acids. The lactobacilli produce several substances that have antimicrobial activity against various microorganisms. They produce lactic acid with resultant pH change. They also produce hydrogen peroxide, which is toxic for some organisms, as well as diacetyl, which has antimicrobial activity. Daeschel (1989) reviewed the antimicrobial substances from lactobacilli (nisin is one such substance) and proprionibacteria. Some show promise against many foodborne pathogens including *Listeria monocytogenes*.

Another review (Stiles and Hastings, 1991), published just 2 years later, demonstrates the amount of research taking place in this field to discover new preservatives. Stiles and Hastings classify the bacteriocins as follows:

- those produced by *Lactococcus* spp.;
- those produced by *Lactobacillus* spp.;
- those produced by *Carnobacterium* spp.;
- those produced by *Leuconostoc* spp.;
- those produced by *Pediococcus* spp.

As Stiles and Hastings point out, beyond that classification, their ability to inhibit microorganisms and the fact that they are all proteins, bacteriocins have little in common. They vary widely in everything else: the spectrum of microorganisms they are effective against, their molecular weight, how they perform, and their biochemical properties.

Some examples of the variability of the properties of bacteriocins are worthwhile. Juven and co-workers (1992) isolated a bacteriocin, designated

LA-147, from a strain of *Lactobacillus acidophilus* found in chicken intestine. LA-147, a protein with a molecular weight of 38.5 kDa, had an extremely narrow spectrum of activity against strains of *L. leichmannii*.

At the other extreme, Barnby-Smith (1992) reviews the colicins from *E. coli*, the lantibiotics (from a variety of microorganisms), which have a broad spectrum of activity wider than most other bacteriocins, and the pediocins (from *Pediococcus* spp.), whose antimicrobial activity spectrum includes *Listeria monocytogenes*. She focuses more on how these bacteriocins might be applied in the future. (At present, only nisin is permitted.) The bacteriocin could be isolated, purified, and added as the pure compound. Or, with the knowledge of which bacteriocins are developed by which strains, particular strains could be added to the food as a starter culture. Bacteriocins show promise in meat preservation.

G. Colorants

The gradual disapproval of many of the synthetic colorants because of questions of their safety is known to all product developers. Equally well known is the importance of color to the appeal of foods.

The hunt by ingredient suppliers for natural colorants to replace those banned has spurred much research to find suitable natural colors, which remain stable under the rigorous conditions frequently found in food manufacture, for example, hard-boiled sweets.

Engel (1979) looks closely at three classes of natural colorants: anthocyanins and anthocyanidins, carotenoids, and betalains. Factors that influence the stability of each class of natural colorant (heat, light, oxidation, presence of metal ions, reactions with other food components, pH effects, etc.) are discussed, with the advantages and disadvantages of each as used in foods.

Anthocyanins are ionic. Consequently, their color is greatly influenced by the pH of food. Furthermore, they are not stable to heat, oxygen, or the presence of sugar and sugar-browning products. Sulfur dioxide bleaches them.

Carotenoids, of which there are over 300, are generally thought of as fat soluble only. However, some, bixin (from annatto, which is used in coloring cheese and baked goods), astaxanthin, and crocin (from crocus blossoms), are water soluble. Their highly unsaturated, conjugated double bond structure provides a range of red to yellow hues. Blending increases this range of hues. However, their very unsaturated nature makes them easily susceptible to oxidation. They are also light sensitive.

Betalains are water-soluble, quaternary ammonium compounds and as such are also ionic in nature and subject to changes in color with changes in pH. They are temperature, oxidation, and light sensitive.

Francis (1981) reviews the anthocyanins, betacyanins (betalains), and yellow/orange pigments (carotenoids and turmeric, which contains the colorant curcumin). Some miscellaneous pigments, Francis notes, are two of insect origin, cochineal and laccainic acid, a red extract produced by the mold *Monascus purpureus* with a long history of use in the Orient, and chlorophyll, used in some specialty pasta products.

Work on the separation and identification of anthocyanins from sweet potatoes is described by Gabriel (1989) and Shi and co-workers (1992a). Shi and co-workers (1992b) studied the stability of anthocyanins in model food systems.

A major problem with natural colorants, in addition to their instability in many food systems, is their safety, which is far from being established, and their scarcity (hence their cost). As Francis (1981) points out, 75,000 hand-picked crocus blossoms are required to obtain a pound of crocin!

An exciting discovery is reviewed by Francis (1992). A new group of anthocyanins substituted on the B-ring were found. This substitution provides greater pH stability and brighter colors. Acylated B-ring substituted anthocyanins are not commercially available, but as Francis notes, they show promise as a new class of stable colorants. They are, however, all from nonfood sources and would require extensive safety testing.

H. A Cautionary Note

Ingredients used in the formulation of food products are in a delicate balance. Each contributes some characteristic to the product. Great care must be taken in understanding what contributions each makes in the product, either singly or in combinations where synergism of actions may play a role. Altering, removing, or substituting any one component upsets the balance in the formulation.

Product developers must understand the purpose and contribution of each ingredient. The desired characteristics that each contributes must be recorded. If they are not, there is a real risk that substitutions with other, perhaps cheaper, ingredients may later be done with disastrous results to product safety, product stability, or product quality.

Chapter 5

Refining the Screening Procedures for the Product

I see you stand like greyhounds in the slips,
Straining upon the start. The game's afoot:
The Life of King Henry The Fifth, Act III, William Shakespeare

Technical members of new product development teams screen new product by the very nature of their work. Limits to what the food plant can and cannot physically do with respect to a finished product are delineated. Purchasing personnel obtain accurate cost figures for ingredients and raw materials as specified by technologists, which may indicate that a price to consumers will be out of line. Engineers recognize limitations of plant equipment and identify modifications that will be required in the plant; production has estimates of labor requirements. Each of the above steps provides a screen for further activities.

With the cooperation of the production staff, developers must establish specifications for raw materials and ingredients. Suitable analytical procedures to test these must be selected. Verification of thermal processes may be required; certainly, hazard analysis critical control point programs that suit the desired level of product integrity are required. Staff must be trained and new skills learned if the new products require new processing technologies and new ingredients.

But marketing still does not know if it has a product that it can sell. Consumers must want this product: the product must fill a need and meet the expectations of consumers.

One of the major requirements of food products is that they taste good. All the sensory characteristics that are associated with a particular food must be met. To ascertain this in a new product, its sensory appeal must be measured. Herein begin problems in measuring something as subjective as sensory appeal.

Technologists, for their part, must still determine how well the stabilizing systems that they designed work in preserving these desirable sensory characteristics and in reducing risks of public health significance. In short, they need to determine the shelf life of their product. Rates at which products can deteriorate are an important consideration in the choice of packaging materials, in requirements of the distribution system, and in label statements.

Much of the trial and error needed must be foreshortened under the competitive pressures to get out as quickly as possible into the marketplace and always within or under budget. With the development of the computer and suitable statistical software, much of the development time and replication of testing can be cut to free up staff, reduce development costs, and speed the achievement of meaningful results.

I. SENSORY TECHNIQUES

Sensory appeal is difficult to measure. Peryam (1990) describes the initially slow and then more rapid growth of the history of the understanding and application of sensory evaluation to product development. Sensory testing must be carried out rigorously, with proper experimental design and correctly selected and trained sensory panelists, to minimize the errors that can creep into any scientific trial. This is why technologists knowledgeable in the techniques should carry these out. Data from organoleptic tests require special statistical skills for analysis and interpretation, which is why sensory measurements should not be left to those who have had no training in the field of sensory analysis. Some populations of tasters, such as children, present unique problems in assessing sensory appeal which is why special skills are required in sensory analysis. It should be apparent that sensory evaluation measurements and their analyses should be carried out by technologists trained to do them properly. At the very least, a company needs some member of the research and development staff familiar enough with the technology to be aware of their own limitations in conducting the tests and of the limitations of the tests themselves.

Sensory technologists must have absolute authority in conducting the tests. Neither project managers, nor technologists in charge of the product, nor presidents (i.e., no one) should be allowed to dominate or disrupt a sensory test. Interference in the conducting of sensory tests is a situation which I have seen occur too often, especially in smaller companies.

There are four superb references, any one or all of which sensory technologists or any person contemplating undertaking sensory evaluation measurements should have as a vademecum readily available for easy reference:

1. Laboratory Methods For Sensory Analysis of Food, written by L. M. Poste, D. A. Mackie, G. Butler, and Elizabeth Larmond (Research Branch, Agriculture Canada, Publication 1864/E, 1991) is a compact handbook written in an easy to assimilate style (Poste et al., 1991).

2. Sensory Evaluation Guide For Testing Food and Beverage Products, prepared by the Sensory Evaluation Division of the Institute of Food Technologists (1981) is a concisely written guide describing the tests that can be used and providing references where one can find further details.
3. Tasting tests carried out at the Leatherhead Food Research Association, in the *Leatherhead Food Research Association Technical Circular* (Williamson, 1981), prepared by Marion Williamson, is a very readable document giving many of the descriptive terms used in sensory analysis and providing, in an appendix, concerns for the safety aspects of tasting panels (expert panels) and the need to have a clearly defined statement of policy on the use of tasting panels.
4. The Bureau de normalization du Quebec (Ministere de l'Industrie, du Commerce et du Tourisme, Gouvernement du Quebec (Quebec Bureau of Standards, Ministry of Industry, Trade and Tourism, Government of Quebec) has published an excellent series of documents covering:
   vocabulary (BNQ 8000-500; 84-03-06),
   general methodology (BNQ 8000-510; 84-03-06),
   sample preparation (BNQ 8000-512; 84-03-05),
   scaling techniques (BNQ 8000-515; 1982-10-08),
   triangular tests (BNQ 8000-517; 84-03-06),
   paired comparisons (BNQ 8000-519; 84-03-06),
   designing a sensory analysis location (BNQ 8000-525; 1982-08-27),
   determination of taste acuity (BNQ 8000-560; 1982-10-08),
   methods for determining flavor profile (BNQ 8000-570;84-03-05).

Two types of sensory evaluation, whose functions must be clearly understood, are objective sensory evaluations and subjective sensory evaluations. Each has quite specific purposes. They cannot be interchanged. Far too often, the purposes for which the sensory tests are meant are either misunderstood or are ignored. Companies will use the results of an objective test as indicating a sensory preference for one product over another. This is wrong and can lead to incorrect decisions in product development that can be disastrous.

A. Objective Testing

Objective sensory evaluation tests are used to get an objective evaluation of some sensory appeal. Other names used for these tests are more descriptive of their purpose: analytical sensory tests, expert panel tests, difference tests. The questions asked of panelists are the following:

- Does a difference exist between the samples?
- How would you rank the samples with respect to the strength of some sensory characteristic?
- How would you describe and rank the sensory characteristics you can identify in a sample (profiling)?

Objective tests are used to determine if a difference exists between products with respect to some sensory quality or between a reference and a test sample. For example, Jeremiah and co-workers (1992) followed changes of chilled pork loin packaged under carbon dioxide and vacuum and stored at $-1.5°C$ for up to 24 weeks. They used consensus profiles. From a pool of 12 highly trained flavor/texture panelists, at least 6 at any one testing profiled each sample individually. In subsequent discussions, a consensus profile was developed. Details of their procedure of analysis can be found in the reference.

Objective testing requires trained panelists. Poste et al. (1991) provide advice on the selection and training of panelists. Like wine tasters, panelists must develop a common vocabulary, so they can describe sensations. They must be trained to recognize different sensations, to be discriminating and to quantify these sensations against recognized standards. Rutledge (1992) describes what she refers to as an accelerated program for the training of descriptive flavor analysts. The program requires up to 9 weeks of training. Novice flavor analysts are coached, guided, and tutored by established analysts, much like an apprenticeship program. In an earlier paper, Rutledge and Hudson (1990) describe a general method for the training of descriptive flavor analysis panels. Powers (1988) describes the use of statistical procedures, such as multivariate analysis, for the screening and training of sensory panelists.

Commonly, many food companies select their panelists from staff. Sensory analysis is, therefore, not their only job. Indeed, sensory analysis may intrude upon the panelist's regular company responsibilities. Sensory specialists must be careful both with selecting panelists, keeping them in top tasting form, and not intruding on workloads so that they are not irritated nor do they feel threatened if normal workloads are tardy. Panelists must like what they are doing and bring no biases to the sensory testing room. Williamson (1981) treats this aspect of selecting panelists at length.

Sensory technologists should keep a large pool of panelists. This will compensate for the usual absenteeism and nonavailability that can occur. Another reason for having a large pool is that not all panelists are suitable for testing all sensory characteristics under study. Sensory specialists should keep profiles of each panelist, listing availability, threshold levels of discrimination, and sensory record in the various testing sessions the panelist has attended (see Powers, 1988).

B. Preference Testing

In preference testing, also called subjective testing or affective testing, panelists are presented with a choice of samples and must state which sample is preferred. The word "preferred" is understood to mean "is most acceptable", "tastes best", "looks best", "would buy", or any other expression indicating greater satisfaction. In the example in Chapter 4, preference testing would certainly have revealed to the tortilla chip manufacturer that the costly packaging improvements did not provide an added-value feature to the consumer.

There are three main variants in the way preference testing can be carried out: the focus group, the central location test, and the in-home test. The first technique is similar to a focus group. A small number of panelists representative of the targeted consumer are asked to test the product and fill in a questionnaire concerning the product. This process will be repeated several times and always with selected consumers.

Or, the new product development team may take product under test to some central location. Here, the team will be assured of a broad cross-section of potential testers. Many companies select church groups, veterans associations, shopping malls (if the surrounding district is appropriate with respect to the target), ethnic clubs (with consideration of who the target consumer is), or social groups for their testing. These groups can be selected from locations all over the country. A carefully prepared questionnaire should be used by the new product team to evaluate consumer preferences.

The above tests, the focus type and the central location tests, provide a degree of control over the product. That is, samples are prepared uniformly and in a controlled fashion. Qualified personnel carry out the interviews or are at least available for interpretation of questions on self-administered questionnaires.

In the third variant, the in-home test, preselected consumers (There is some control of the consumer.) are sent coded samples of product. These are to be prepared at home. There is no control of the preparation of the samples, nor is there control of the environment in which the test is carried out. How well were the instructions followed with all the distractions that are possible in the home? Were the time, place, and other circumstances in the home conducive to a "good" test? Consumers then fill out and return the accompanying questionnaire, or they are interviewed by telephone.

The characteristics of the preference testing techniques are compared in Table 5.

Who should be panelists in any of the above preference tests? Under no circumstances should trained (objective) panels be used for preference (subjective) testing. It is not that these people do not have preferences; they are simply not at all representative of consumers. Their preferences have no marketing validity with respect to the population.

Two examples that I have experienced will suffice to illustrate this point. The first was already described in Chapter 4 and involved the manufacturer of tortilla chips, whose expert panel determined that an incipient taste of oxidative rancidity was offensive in an established tortilla chip product. The expert panel was proven wrong: consumers preferred that flavor.

The other example involved a company determined to upgrade the appearance of a canned pour-over sauce. (There was no consumer-initiated demand for improvement in either this incident or the one with the tortilla manufacturer.) The sauce was improved in color, according to its expert sensory panel; it was a richer color. It no longer showed evidence of syneresis and was smoother looking. These improvements were clearly identified by its expert

**TABLE 5. Preference Testing: Three Main Variants
of Each, with Characteristics**

Test Variant	Characteristics
Focus group	• 8 to 12 participants • respondents can be carefully selected • can be easily repeated • fairly economical to conduct • requires professional leader • control over product good • results quickly obtained • qualitative data only • data nonprojectable
Central location test	• involves larger groups; i.e., social clubs, church groups, etc. • poorer selectivity of respondents • can be easily repeated • somewhat more expensive to conduct • requires well-prepared questionnaire • control over product excellent • results quickly obtained • quantitative and qualitative data obtained • data projectable with caution
In-home test	• can involve several hundred respondents • good selectivity of respondents • usually a one-time test • more expensive to conduct than preceding • requires intensive follow-up with well-prepared questionnaire • poor control over product • results slowly obtained • quantitative and qualitative data obtained • data projectable with caution

panel as better. They were accomplished by slight recipe modification — an economic advantage, since less hydrolyzed vegetable protein was used — and mostly by a new and expensive, retorting operation. With fanfare, the new improved product was launched, only to be received by a deluge of consumer complaints. The gist of the complaints was that the product now looked glossy and ersatz; it no longer had that homemade "just like Momma's" appearance.

Unfortunately, many companies still use their expert panels to get data on the acceptability of a product. This is wrong.

Nor is it wise (but it is done in larger companies) to use employees, either from the plant floor or from the offices, for preference testing. These people are not members of an expert panel, but their knowledge of what taste image their company has may bias their opinions. That is, they are not tasting a product; they are tasting a brand that they know and are very familiar with. They are not typical consumers: they have a bias. They are experts. In my experience, comments such as "... this isn't our company's flavor" or "... I wouldn't want our company to put out a product like this" will prevail and confound whatever results are obtained. These testers know nothing of the company's objectives in this test. The company may be working on developing another brand in a very different marketing niche. Where, however, companies have a strong brand image, these same panelists may have a better idea of what the brand can carry than does management. The point must be made strongly that company workers, particularly the manufacturing workers, can bring a very definite brand bias to preference sensory testing.

MacFie (1990) and Gutteridge (1990) discuss sensory techniques affecting consumers. MacFie describes characteristics affecting consumers' choices and explains "free choice profiling" and preference mapping. Gutteridge (1990) discusses the technique of repertory elicitation with statistical treatment (REST®, Mathematical Market Research, Ltd., Oxford, U.K.; see also Thomson, 1989) as a method for finding the most appropriate market niche for products.

C. Considerations in Sensory Analysis

Before sensory testing begins, there must be a clear notion of what is required from the test. Does the product development technical team want to determine whether they have produced the best formulation of a product with respect to whatever criteria? Or do they want to determine whether this particular formulated company product is as good as or better than the competition's product? That is, are the results of technical product development under investigation, or are marketing personnel trying to understand the product in its full marketplace context where branding may be a factor? Answers to these questions about product characteristics have a decided impact on how the tests will be conducted and who will do the testing.

Martin (1990) found product characteristics could be broken into three components:

- physical characteristics of a product, which are well recognized by the consumer as poor, acceptable, or high quality characteristics; the technologist is trying to formulate only these high quality (ideal) attributes into the product;
- image characteristics, which Martin recognizes as most dominant in perfumes, fragrances, and tobacco products but which are not unknown in food products, e.g., liqueurs, liqueur-flavored instant coffees, exotic gourmet sauces;
- a combination or interaction of physical and image characteristics.

The interaction component is interesting, in that it brings up an interesting problem. Martin (1990) describes the interaction as follows: when testing similar products "blind" (for example, various brands of beer) in sensory tests, consumers place the product ideal (the "best" product) some distance away from the test product and with other unbranded competitive products. When, however, *branded* products are evaluated (that is, the test is no longer conducted "blind"), all products are ranked closer to the consumer's ideal product. Branding is an important factor to be considered in sensory testing.

1. To Test Blind or Not?

The foregoing raises an interesting issue. Should a product be tested branded or unbranded? After all, the product will have to face other similar products in the marketplace eventually, and most of these will be branded. Brands have uniqueness: they communicate images, a product persona. Therein lies a brand's value. A brand is comfort, known values, security to consumers.

There is overwhelming evidence that branding does influence tasters in ranking similar products and in picking a preferred sample. Martin (1990) provides evidence of vastly different assessments of various characteristics of ciders, beers, and chocolate confectionery when the products were tested branded or blind. In tests on beer, Martin found that in one instance, one beer, when identified, was rated more poorly than when it was tested blind.

Moskowitz and co-workers (1981) describe an interesting analysis of consumer perception, magnitude estimation scaling, carried out on chocolate candy bars. Their results demonstrate that branding encourages a product's acceptability. Further, they determined that for some products, it could be branding and not a product's quality that lifted a product's acceptability. But one must be aware of a difference here: branding influences acceptability, image, the "persona" of a product. It does not influence the evaluation of objective characteristics posed by questions, such as, "Is product A smoother than product B? or sweeter? or sourer?"

Schutz (1988), Scriven et al. (1989), and Gains and Thomson (1990) describe sensory techniques to evaluate consumer attitudes to foods and determine the circumstances under which consumers would use or serve particular foods (contextual analysis). Such techniques are excellent tools to guide marketing personnel in determining market niches for products.

The questions must be repeated: What is the purpose of the sensory test? Does the company want to know whether it has the best formulated product? Or does the company want to know whether it has a product that is preferred over the competition's product? This is a technological capability question vs. a marketing capability question: technology vs. psychology. As Martin (1990) aptly put it:

> It may not be necessary to develop a clearly superior product in sensory terms, if reputation can sufficiently enhance one which is the equal of the competition.

Hardy (1991) is more adamant and maintains that superior tasting food and beverage products are no surety of a loyal consumer base.

2. What About the Sensory Testers?

Through their application of magnitude estimation scaling to the candy preference survey, Moskowitz and his colleagues (1981) note — at least for chocolate candies — that there could be age differences and sex differences in the response to branding. That is, the respondent's age and sex influence how a product is perceived or accepted. Clearly, this suggests that great care is required in the selection of panelists.

Yet, there is evidence that when faced with unbranded popular products, most tasters cannot distinguish between the unbranded products nor even successfully choose their favorite brand. For example, Hardy (1991) found that most tasters not only could not distinguish consistently between competitive products but did not improve with experience without formal feedback (i.e., training). Sensory testers must decide what is the purpose of the test and consider whether the biases that branding may introduce will adversely affect this purpose.

Ideally, sensory panelists should be representative of those for whom the new product is targeted. Ideally too, all geographic areas where the target consumer can be found should be represented in consumer research tests. However, the ideal can seldom be achieved without assistance.

Consequently, food companies enlist the aid of product testing/consumer research companies. These groups keep extensive files of potential panelists categorized by age, sex, ethnic background, economic status, and other characteristics that may be important. McDermott (1990) discusses recruiting for sensory testing, problems encountered with setting specifications on recruitment, and how these involve cost considerations.

Money limitations may provide constraints on both the number of subjective tests that can be carried out and the size of the panels used in them. When to test, how to test, what to test, and how big a test to conduct are decisions companies undertaking product development must be prepared to make.

Products designed for young children present unique problems. There is first the difficulty of working with young children, communicating with them, and determining how to measure their preferences. Skilled panel leaders are required, and it is virtually a necessity that the relationship be one-on-one, that is, one leader to one child. Obviously such testings can be very expensive because of the extra skills required. Then there are decisions concerning the test itself. How big should the test be with children? How often can the test be conducted?

Kroll (1990) describes another difficulty with testing children: what scaling system to use. She tested children using one-on-one interviewing, a self-administered questionnaire, and three different types of rating scales. She found that while all three scales discriminated at the 10% level, a simple in-house scale used by Peryam and Kroll (Peryam & Kroll, Chicago, IL) was better than a traditional hedonic scale and (surprisingly) better than a face scale.

II. SHELF LIFE TESTING

A product's shelf life is a verification of the stabilizing systems designed into the product. Food companies cannot release new products into a market, especially a test market, without knowing how stable that product will prove to be. That is, the company must have some idea of what the product's shelf life is. This seemingly simple requirement, to provide a shelf life estimate, is fraught with all sorts of difficulties. Curiale (1991) discusses some of these, with particular reference to microbiological shelf life. His and other points to consider in shelf life determinations follow.

A. Selecting Criteria to Assess Shelf Life

First, some criterion that changes with time and that is appropriate for the product must be selected. Changes of this criterion can be measured with respect to time. Microbiological criteria may be chosen (Curiale, 1991). For example, with chilled foods, total plate counts, psychrophilic counts, or counts of specific microorganisms of public health or economic significance may be monitored for the estimation of shelf life. The loss of some nutrient such as vitamin C might be chosen. Or the loss of some functional property of the product, such as its ability to whip, to color, to foam, or to leaven might as easily be picked. The progressive gain of some undesirable textural change, hardening or softening or staling, can be the criterion for assessing shelf life. Loss of crispness in potato chips can be the mark of a loss of quality, as can the development of graininess in a fondant or in an ice cream.

Selection of any of the above criteria for stability presents different degrees of complexity for developers. It is comparatively simple to follow the destruction of a nutrient like vitamin C in a food. But if that new food is not an important or even significant source of vitamin C, then vitamin C is not, in all likelihood, a useful criterion for monitoring shelf life. If the loss of vitamin C correlates closely to the loss of a major quality characteristic of the new food product that is difficult to measure, then vitamin C is a good standard.

Most foods are complex systems of components and cannot usually be judged for quality on the basis of one characteristic. Rarely is only one quality characteristic of a food adversely affected during its shelf life. It is more likely that color, texture, and flavor will all degrade over time. And frustratingly, they will degrade at different rates. Therefore, choosing the correct criterion or criteria to follow during the determination of shelf life stability becomes very important.

Second, there must be some judgment respecting how much loss of quality characteristics can be accepted (and by whom)? Loss of some functional property raises the question: how much loss of functional property can be accepted before spoilage is declared? If the criterion is color, then how much color loss is acceptable? The loss of some nutrient cannot be seen or tasted, but where a label declaration has been made for that nutrient, unacceptability acquires a new meaning, i.e., a label violation. Losses of functionality, color, flavor, or texture are assessed by consumers. The result can be dissatisfied consumers.

What constitutes an acceptable degree of instability? Is it the loss of 60%, 50%, or whatever percentage of the vitamin C content or of the redness or crispness of the product? Or is it when the microbiological plate counts reach a particular level? Developers may speak in terms of "high quality shelf life" or "acceptable quality shelf life" or "useful storage life", but these terms do not really have any meaning except to other developers. They have no meaning to consumers.

A final decision to be made is this: under what conditions will the shelf life test be carried out? If ideal temperature conditions of frozen storage for a frozen new food product, or ideal temperature conditions of refrigerated storage for a refrigerated new food product, or even the beloved "ntp" (normal temperature and pressure) of physical chemists is used for thermally processed foods, the resultant shelf life that is determined will probably bear no likeness to what will happen in the real world of consumers. For frozen or chilled foods, this world may include the following:

- the temperature changes encountered in factory warehousing,
- temperature changes during transport and transfer to a wholesaler's or a retail chain's warehouse, storage in this warehouse, subsequent transport, and transfer to a retail outlet (Slight, 1980),
- perhaps a short period of temporary storage in the retail outlet, then shelf display in refrigerated cabinets stocked improperly and frequently improperly maintained,
- storage in vending machines, which for frozen or refrigerated goods, can be very damaging (Light et al., 1987),
- further temperature abuse from transport and in-house handling by consumers.

This, then, is the dilemma to be faced in determining the shelf life of a product. Is the product to be tested under nonabusive conditions? That is, are recommended conditions of storage being followed throughout? Or is it under abusive conditions, such as are encountered in the distribution chain?

Calculation of the shelf life has very serious economic impact, especially where statements such as "best before …" or "best quality before …" are required by label legislation. If too short a shelf life is stated on the label, retailers will return product past its stated expiry date, believing the product to be either deteriorated or unhealthy. This constitutes a loss to the manufacturer. If, on the other hand, an exaggeratedly long shelf life is projected, a large number of consumer complaints may arise from failed product.

A rough rule of thumb that is frequently used, although many may deny that they do, is the two-thirds rule. The rule works as follows: shelf life tests show that a particular packaged product has a good quality shelf life of 90 days at refrigerator temperature. According to the two-thirds rule, the shelf life is stated as 60 days. I have never found any practical nor scientific justification for this, but I have found, nevertheless, that it is used.

B. Selecting Conditions for the Test

Three factors determine the quality shelf life of a food product:

1. the preservative systems designed into the food product, which are protected by the packaging material selected for this ability,
2. the physical abuse that food handlers — the people involved in the warehousing, distribution, retailing, purchasing, and in-home storage of the food — can give the package of food,
3. the environmental abuse that the product and its package will encounter from manufacturing and packaging until it is opened by the consumer and consumed.

A fourth possible factor has been omitted: the microbiological load on the food after processing and on the food as it is packaged. It is assumed that good manufacturing practices and a sound HACCP program were in place during manufacturing. However, for chilled foods and minimally processed foods, such an assumption is not valid. The initial microbiological load is an important factor in an acceptable shelf life. A sound HACCP program is critical to these foods.

There can, therefore, from factory door to consumer's table, be several factors that affect a product's shelf life adversely:

- interactions between the various chemical components of the food that can alter quality. These were anticipated, and developers designed stabilizing systems to prevent them;
- interactions between the food and its package that can damage the integrity of the package and alter the quality. These, too, were anticipated;
- adverse effects of the environment (temperature changes, relative humidity changes, irradiation), to which the package is subjected during these interactions. These could not be anticipated.

If that were not enough, all these changes can be confounded by abusive treatment.

For example, temperature influences the rate of chemical reactions, as every secondary school student knows. The familiar Q_{10} tells that for every 10°C increase in temperature, one can anticipate a two- to fourfold increase in the rate of a chemical reaction. This influence of temperature on rates can be seen in the work of Labuza and Riboh (1982), who reviewed the influence that abusive temperature treatment can have on Arrhenius kinetics with respect to nutrient losses used as the predictor of shelf life of foods.

Temperature changes will cause phase changes in foods. Gels and emulsions will break down; ice will thaw, or ice crystals can grow and damage the structural integrity of soft foods. Texture or tackiness of foods will alter (Slade and Levine, 1991; Goff, 1992).

Temperature changes affect biological reactions. Such changes can alter growth rates and growth patterns of microorganisms and the activity of enzymes (Thorne, 1978; Williams, 1978).

Temperature changes can occur in warehouses from doors being opened and closed during receiving and shipping. Heaters in warehouses can warm product stacked too near them. Changes occur during transportation in refrigerated vehicles with improperly set or maintained thermostats. Retail cabinets with their defrost cycles impose further temperature changes. Humidity changes closely associated with temperature changes can cause sweating of packages, resulting in rusting of metal containers or label damage outside the package. Incident light striking exposed cartons or containers on ship decks can cause large temperature changes in foods.

Product development technologists have traditionally used three general approaches to determine shelf life:

- static tests, in which the product is stored under a given set of environmental conditions selected as most representative of the conditions to which the product will be subjected;
- accelerated tests, in which the product is stored under a range of some environmental variable (for example, temperature);
- use/abuse tests, in which the product is cycled through some environmental variables.

At intervals through these tests, samples are taken and subjected to sensory, chemical, and microbiological assessment.

1.   Static Tests

Static tests have obvious shortcomings. First, it takes too long for noticeable changes to occur in a food; second, because of this fault, the tests are too costly to undertake given the paucity of information they provide to investigators. A static test can be likened to a one-point viscosity measurement or a one-point moisture sorption curve: it tells nothing of the behavior of the product under other stresses. It provides no kinetic data.

An example of the use of a static test is provided by Okoli and Ezenweke (1990), who developed a new product, pawpaw juice, for consumption in Nigeria. Such a product was unknown there, although freshly sliced and peeled pawpaw was well known and accepted. Their storage conditions were $30 \pm 2°C$ for the test bottles of pawpaw juice and $10°C$ for the comparison controls. Samples were subjected to both sensory analysis and physical and chemical analyses at intervals over 80 weeks. For commercial development, this is much too long and would certainly frustrate the marketing program of most companies.

2. Accelerated Tests

Accelerated tests are preferred by most researchers, since such tests provide much more information about products and the kinetics of their deterioration. Factors to be considered in accelerated tests are well documented by Labuza and Schmidl (1985). A range of conditions of some environmental variable (such as temperature) that could be anticipated in distribution is carefully chosen. The product (in its container) is stored under each of the temperature conditions and analyzed at intervals for the loss of some valued quality characteristic. If the variable chosen is temperature, a simple application of the Arrhenius equation allows researchers to demonstrate graphically the relationship between temperature and time in days until an undesirable degree of loss of quality occurs. Then researchers can calculate the number of days of good shelf life to be expected if one assumes good storage conditions.

Conditions of the accelerated test must be selected with a degree of care (Labuza and Schmidl, 1985). If one considers only temperature as the environmental variable, one must remember that ice can thaw, and conversely, water can freeze; fat can melt or solidify; suspensions (emulsions) can degrade; and the rheological properties can change drastically. These changes can seriously skew data obtained by this sort of kinetic modeling.

Obviously, there is a limit to increasing the storage temperature of frozen foods, but this limit might be lower than one might think. Water can still be liquid in a food system well below its freezing point as pure water. One is no longer studying the behavior of a frozen product but rather a "liquid" product.

Ideally, the variable chosen for accelerating should not alter the normal or anticipated path of spoilage, i.e., the path of spoilage to be expected in the normal, nonabusive condition. The purpose of an accelerated test is primarily to determine the shelf life of a packaged new food product under normal marketing conditions whatever these may be and only secondarily to study reaction kinetics of its deterioration. Labuza and Schmidl (1985) provide a table of recommended storage temperatures for frozen foods (–5 to –40°C for the control), for dry and intermediate moisture foods (0°C for the control to 45°C), and for thermally processed foods (5°C for the control to 40°C).

That said, it must also be remembered that deteriorative reactions have different rates. The temperature sensitivities of some quality characteristics of foods vary widely. That is, at temperature T°C (see Chapter 4, Figure 12), attribute 2 (which could, for example, be flavor) is the limiting quality characteristic. At temperature $(T + 5)°C$, attribute 1 (for example, color) may be the limiting quality characteristic to a good quality shelf life, while at $(T - 5)°C$, attribute 3 limits acceptability. Again, reference to Labuza and Schmidl (1985) will provide guidance in determining both storage times and intervals between samplings.

Temperature is not the only accelerating factor that can be used, but it is the usual one. Chuzel and Zakhia (1991) used adsorption isotherms at three different temperatures to derive an equation describing the shelf life of gari in terms of:

- its equilibrium moisture content,
- its safe storage moisture content,
- its initial moisture content when packaged,
- the permeability of the package and its surface area,
- the fill weight of the package,
- the slope of the product isotherm.

Their suggestions for further work are to include the initial microorganism load and quality characteristics such as color, flavor, and crispness. Gari is a semolina prepared from cassava, which has been fermented, cooked, and dried and is popular in West Africa and in Brazil (farinha de mandioca).

3. Use/Abuse Tests

Use/abuse tests are different from either static or accelerated tests. They are included here because many technologists do use them as a tool in assessing the shelf life of the food and its package as a unit. They are as varied as an imaginative mind can make them.

Frozen food developers commonly cycle new frozen food through the temperature range of −10 to +20°F. This range duplicates the freeze-thaw cycles of frost-free frozen food cabinets. Cycles are set to correspond to those that could be anticipated under store conditions. Some developers purchase freezer display cabinets, stack them in the manner seen in most supermarkets (with a good portion of the product above the recommended level), and thus simulate real supermarket conditions.

In one instance of which I am aware, a company's frozen product was stored in freezer lockers set to cycle at the temperatures that their experience had taught them would be encountered from factory warehouse to frozen wholesale warehouse through to the retailer. In between these stages, product was removed to simulate the transfer from storage to ambient temperature to frozen transport to dock transfer at the next stage and so on.

In another example of a use/abuse test, a pallet of cased product is dispatched on a journey around the country by truck and by train. The rigors of transportation and the influence of temperature and humidity changes due to weather on both the product and the package can be studied. On the pallet's return to the plant, many evaluations can be made concerning the effectiveness of packaging and the condition of the product. Application of this data to shelf life calculations can be limited, but information on what one can expect of the product and its package in handling and distribution can be useful in protecting the validity of one's shelf life statement. Perhaps this was the origin of the two-thirds rule?

Cardoso and Labuza (1983) studied the moisture gain and loss of egg noodles packaged in paperboard, polypropylene, and polyethylene, typical packaging materials for this product. The products were cycled under controlled but varying conditions of temperature (30 to 45°C) and relative humidity (11 to 85%). From their data, they developed a kinetic model to predict moisture transfer, an important factor in product stability.

Porter (1981) discusses the unique problems the military has in shelf life prediction under noncontrolled environments, a predicament the military knows only too well when shipping food from controlled temperature facilities to Arctic bases or to arid desert conditions, such as in the recent Iraqi conflict. However, food manufacturers face many of these same problems in exporting their products to foreign countries with vastly different environmental conditions, as well as warehousing and shipping conditions. Temperatures in container shipments can reach levels that can be very stressful to a good quality shelf life.

Three factors, it was stated earlier, determine the shelf life of a food product:

- its preservative system and the package protecting this,
- abusive handling treatment from factory to consumer,
- the environment which the food and its package encounters throughout its shelf life.

Static tests and accelerated tests can challenge the first and simulate partially the last, the environment. Use/abuse tests answer to all three factors. It is difficult to predict any estimation of shelf life with absolute certainty from such data, and no tests can be undertaken to simulate all the circumstances that may befall a product from manufacturer to consumer.

Abusive or even unusual treatment that a food might encounter during storage or transportation cannot be anticipated. No use/abuse test can simulate all the stress that may be heaped on the food product and its protective packaging. How can one duplicate the military mind that pinholed film-wrapped food packages so that they would fit better into the ration cartons? Or how can the damage done to the package and ultimately to the product itself by retail shelf stockers who slash open food containers as they open cartons or practice basketball shots with frozen chicken be simulated?

These abuses can be eliminated only by education of the food handlers in the safe handling of food products. Food manufacturers have relied on retailers to do this, but unfortunately many retailers, with part-time, temporary help and facing very narrow margins of profitability, rarely can undertake to train staff. Nevertheless, abusive treatment of food products can only be minimized by training and education.

Transport of product can result in abusive treatment. Transport of product in containers where the container is exposed to the hot sun during a trans-Atlantic crossing has been already mentioned. Temperatures inside such containers can increase to well over 120°F.

Vibration caused by transport results in product changes. Positions in the hold of a ship in relation to the engines can produce a gentle vibration that can destabilize some food suspensions. Surface transport, with its gentle, rocking action, has been known to produce subtle and in one case unexpected but desirable changes in a chocolate product. In this instance, a chocolate couverture supplier shipped bulk chocolate by tank car from its factory to a customer several thousand miles away. Demand was so high that it was felt advantageous to all parties to produce this product closer to the customer in a new facility. This step turned into a mild disaster when the customer complained over the loss in quality of the product, although the satellite plant rigorously followed the parent plant's procedures for making the couverture. Investigations revealed that the extra conching action caused by rocking of the tank cars developed the better flavor that the customer preferred.

C. Guidelines to Determining Shelf Life

Determining a product's shelf life can be very difficult. Shelf life is an important characteristic of a food product that may be required by legislation or by contract between a co-packer and a buyer or insisted upon by a retailer. Consumers, at the end of the distribution chain, certainly expect food to have a good quality life until it is consumed, no matter how long and under what potentially abusive conditions it was kept. At the starting end of the chain, manufacturers of sensitive food products, chilled foods for example, are at the mercy of distribution and warehousing companies and depend on them to store and handle products under nonabusive conditions.

Manufacturers must also depend on retailers to practice a stock rotation system in their warehouses. It is also unfortunate for manufacturers of sensitive products that many retail food chains misinterpret the product statement, which reads that a particular product has, for example, a 40-day refrigerated shelf life. They assume that a 40-day refrigerated shelf life starts when they decide to stock their shelves or to plan a promotion. Even when the product has been beyond the expiry date, few manufacturers, especially small manufacturers, have refused to take back spoiled product for fear of losing future orders. These are the facts of life that developers determining shelf life must live with.

Estimating shelf life is a guessing game, a description that would irritate many scientists who model kinetics of spoilage reactions; it is, nevertheless, true. The kindest that can be said is that the stated shelf life stamped on any product has a strong element of guess in it, along with the element of hard data that food scientists can add. Neither consumers nor, for that matter, retailers know what previous treatment a product has had when they read the "best before" date. As data on the deterioration of foods are collected and knowledge of growth mechanisms of microorganisms in foods progresses, this guesstimate can be refined.

The following points should be considered:

1. There are no tests that can be relied upon absolutely to allow one to predict the shelf life of a given food product. *All that such tests can do is provide an approximation.* Experience with similar products in the same product category (for example, chilled foods) can help provide some initial estimates. Data in the scientific and technical literature can help refine these initial estimates. Audits of a similar competitor's products drawn from the retail showcases will provide more data to complement one's own findings. Here too, the complaint records of a company can provide information on a product's stability that may be applied to another product in development. This marks another reason for documenting the complaint files. But one should not waste one's development time looking for the perfect test.

2. Shelf life tests ideally should be carried out only on finished product manufactured on the line (and equipment) to be used in plant production and packaged in the container that will be placed on the shelves for consumers. Product prepared in test kitchens or pilot plants does not simulate the product prepared in a plant at the height of the packing season with adjacent packing lines running products capable of being a source of cross-contamination. Bailey (1988) discusses these problems of scale-up from the pilot plant to the production floor and describes attempts using predictive techniques to minimize the discrepancy between the pilot plant and the manufacturing plant.

   The following examples illustrate the fallacy of relying heavily on test kitchen samples for reliable shelf life data. A manufacturer wanted to develop a line of chilled semiconserved prepared vegetables. Shelf life of the products was determined on pilot-plant-prepared samples made from raw material purchased on the retail market. A shelf life based on several batches of such samples indicated an excellent chilled shelf life exceeding 30 days. In full plant production, a shelf life of 10 to 15 days was the norm. The microbiological load between the samples prepared in the well-kept, sanitary test pilot plant and those prepared in the plant factory under full production with field-grown produce was very significantly different. In another example, the unavailability of the desired container from a cannister supplier and the pressure exerted by marketing "to get on with it", led researchers to substitute a different cannister with plastic ends rather than the desired metal ones for a spiced and herbed bread crumb mix. It was even claimed by the supplier that the plastic substitute would provide a more rigorous test "since plastic breathed". Tests were successfully conducted, and shelf life determined. In full production using the cannisters with the metallic ends, consumer complaints began to pour in. Off-flavors were noted. Rusting was observed. Constituents (particularly citral) in the spice and herb blend reacted with the metal ends to cause a breakdown of the flavor and initiate detinning.

Scale-up from the test kitchen to the pilot plant to the factory floor has always produced changes, some subtle and others not so subtle, in a product. As stated earlier, the manner in which heat is applied is different; stirring action is different, and therefore heat transfer can be altered; pumping action may be different, and the distances the product has to be pumped in the real world of the factory may alter its temperature, its rheological properties, and hence the shear stress the product receives.

Factory-produced product is different from product prepared in the pilot plant. The two products should not be expected to have the same storage stabilities.

3. Once one has estimated the shelf life of a product, any change in the recipe, in the suppliers of the ingredients, in the water treatment system in the plant, or in the water used in batch preparation occasioned by plant relocation, or any other change can possibly alter the acceptable quality shelf life of a food product. For example, the mineral content of plant waters can have a profound effect on the flavor of a product, not only immediately but over a period of time. In another example, the flavor of marinated, fried peppers was adversely affected when the supplier of the frying oil, a fractionated peanut oil, changed from one antioxidant to another one.

4. Extrapolation of data obtained from accelerated storage tests should be treated warily. Extrapolation of data for chilled foods, in particular, can be misleading if due care is not given to the types of microorganisms contributing to the spoilage: spoilage can be overt with obvious indications of slime, off-odors, loss of color, etc., or it can be covert with no obvious outward signs of spoilage but with undesirable increases of microorganisms of public health significance.

5. Acceptance by the consumer, or lack of acceptance, is the ultimate determinant of a product's shelf life. Nevertheless, in basing shelf life on sensory analysis by consumers — that is, on taste — their interpretation with respect to spoilage and their ability to detect this varies considerably. Dethmers (1979) discusses the use of sensory panel

evaluation for failure criteria for open-dating. Curiale (1991) and Beauchamp (1990) both point out some of the shortcomings that can arise in the use of sensory panels for shelf life determinations: Beauchamp, in particular, cites intraindividual and interindividual variations that can confound the use of sensory panels as an evaluation tool. A clear understanding of the criteria to be used to assess the end of acceptable (or high quality) shelf life must be established.

6. Determining the shelf life of products involves measuring the differences over time between control samples and test samples that are subjected to some stress. Wolfe (1979) discusses the advantages and disadvantages of various types of reference standards, especially with respect to sensory studies, but which are equally valid for shelf life studies. See also Labuza and Schmidl (1985) for recommended storage temperatures for control samples.

D. Advances in Shelf Life Considerations

Some exciting new thinking is emerging in the study of the kinetics of food deterioration and in the development of predictive techniques in microbiology. Instead of observing a food spoil over time, could one, from a knowledge of spoilage mechanics and a product's composition, predict with accuracy its expected shelf life?

There has been a need for predictive techniques in microbiology as food manufacturers provide more added value to food products. The result of added value, as Williams and co-workers (1992) see it, is that traditional paths of food spoilage and intoxication have been altered. To alleviate this problem, predictive techniques to control both quality (synonymous here with shelf life) and safety are needed.

The use of predictive models is not new. For example, the botulinum cook established for the safe thermal processing of low acid, high pH foods is an application of a model, an inactivation model, of the destruction of spores of *Clostridium botulinum* by heat (Gould, 1989). The value of predictive models of microbial spoilage is the ability to apply a model to many types of food and, as data accumulates, to many types of stabilizing factors designed into these foods. Rapid predictions of both quality and safety could be made with a greatly reduced need for testing and, therefore, at reduced costs (Roberts, 1989).

Roberts (1989) describes two types of predictive models, the probabilistic model and the kinetic model. Each serves a different purpose. Probabilistic models, as their name suggests, predict the probability of an event, for example toxin development in a food, but provide no information about how quickly the toxin develops or the amount. Roberts (1989) describes such a mathematical model with the ability to predict botulinal toxin production as a function of several stabilizing factors, i.e. salt, nitrite, thermal treatment, storage temperature of the food, and the presence or absence of preservatives (see also Roberts, 1990).

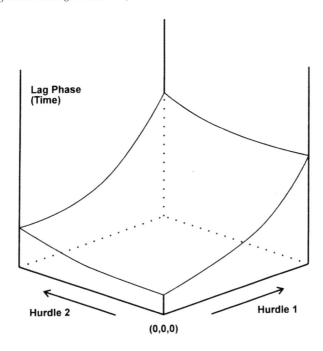

FIGURE 13. Stylized representation of duration of lag phase influenced by various concentrations of two hurdles.

Kinetic models predict the rate of growth of microorganisms and therefore are useful for anticipating time to microbial spoilage or time to growth of critical numbers of food-intoxicating microorganisms. Roberts (1989) discusses several examples of kinetic modeling in his paper. Figure 13 is a hypothetical depiction, which shows three-dimensionally how the lag phase of a microbial growth curve can be extended by the use of concentrations of two different hurdles. This demonstrates the sort of information that can be generated by predictive modeling techniques. This visual presentation is valuable for technologists to design stabilizing systems for new products. It would obviously be advantageous to select concentrations of "hurdle 1" and "hurdle 2" that extend the length of the lag phase of microbial growth (vertical axis), if these are consistent with other quality factors. Likewise, one could determine which stabilizing system causes the slowest rate of microbial growth, which would again contribute to the product's stability. With the aid of predictive models, technologists could design hurdles into the product or its process, from which the length of the lag phase or the rate of growth could be predicted.

The Arrhenius equation has been used for both assessing nutrient losses with temperature (Labuza and Riboh, 1982) and microbiological growth with temperature (Gibbs and Williams, 1990). The Ratkowski square root equation has also been used:

$$r^{1/2} = b(T - T_0)$$

where   r   = growth rate constant
       b   = regression coefficient
      T   = storage temperature (K)
      T_0 = a temperature at which the growth rate is zero

Gibbs and Williams (1990) describe the use of this equation for plotting the growth of *Yersinia enterocolitica*.

However, as Williams et al. (1992) point out, today's new foods can spoil not only due to temperature variations. Low salt, or low acid, or nitrite-reduced, or sulfite-free, or sugar-free variants of food products have multiplied the number of deteriorative routes of conventional products. As multiparameter stabilizing systems are used to preserve these foods, new risks are introduced. There is an increased need to understand the mechanisms of these systems and to be able to predict quality changes, as well as microbial activity, in a complex matrix of variables (see, for example, Chuzel and Zakhia, 1991).

In his review, Buchanan (1993) attributes the growth in the interest in mathematical modeling to three major factors:

- easy access to powerful, number-crunching computers and the software programs to process the data;
- a growing desire of consumers for minimally processed, just-like-fresh foods; this usually requires that the food be in the chilled foods category;
- the need to organize quantitatively and in a systematic fashion the wealth of microbiological data on the vast array of foods potentially at risk as either economic or public health hazards.

Buchanan (1993) discusses and describes various classifications of models. Both he and Gibbs and Williams (1990) provide examples of the use of Arrhenius models and Ratkowski and Gompertzian equations for growth modeling.

Walker and Jones (1992) describe predictive microbiology in general and explain the coordinated research programs that are currently going on in the U.K. This program, under the direction of the Ministry of Agriculture, Fisheries and Food has a host of participating scientists at various laboratories:

- Agricultural and Food Research Council, Institute of Food Research
- Campden Food and Drink Research Association
- Flour Milling and Baking Research Association
- Leatherhead Food Research Association
- Unilever Research
- University of Bath
- University of Wales College of Cardiff.

In addition, private food companies are invited to contribute data. Campden Food and Drink Research Association will house this U.K. Predictive Food Microbiology Database. It will include three units: a database of usable data, a models base of approved modeling programs, and an expert system to allow nonexpert users access to the models, i.e. an intelligent interface. When the system is complete (expected 1994), product developers throughout the U.K. with modems and computers will be able to use the models for predictive microbiology and conceivably both improve the accuracy of determinations of shelf life and reduce the number of trials. Such a database resource would be useful to all in product development.

The same kinetic studies and predictive modeling techniques have been applied to measuring the deteriorative rates for food quality or specific food components. Labuza (1980) used water activity and temperature to measure quality losses in foods and assess water activity's effect on reaction rates. Rockland and Nishi (1980) also used water activity as a variable affecting food product stability. Andrieu and co-workers (1985) used an Oswin type relation ($X = f\{a_w,T\}$) to model pasta drying. Norback (1980) discusses modeling in general for optimization of food processes.

Lund (1983) cites three major reasons for generating models to describe food processes: first, models permit developers the opportunity to optimize processes with the minimum amount of costly trial and error; second, models provide better understanding of processes, which can lead to better processes and safer new products; and third, models permit better prediction of shelf life and quality changes in foods.

Description of modeling techniques and applications to food quality deterioration can be found in Lenz and Lund (1980), Hill and Grieger-Block (1980), and Saguy and Karel (1980).

III. PRODUCT INTEGRITY

The safety of products with respect to hazards of public health significance as well as to hazards of economic significance can be established only by designing safety into the product from the start of development. This concept, designing quality and safety into products, has been well established in instrument manufacturing. Mayo (1986) sees the concept of quality by design at AT&T applying equally to products and to services. He stresses four elements of the program:

- design to the correct requirements of the customer. If the customer's needs have not been clearly identified, product design will be faulty;
- design using the right technology. This has an impact on both the cost of the product and the customer's satisfaction with the product;

- design for manufacturability. As Mayo states it, the product and its process should be designed "... to be insensitive to 'noise' such as conditions of customer use, drift in components, or variations in the manufacturing environment";
- design for reliability.

Coincident with development's progress, feedback mechanisms should continually monitor and analyze the product with respect to quality and safety. This is Mayo's design for reliability.

Huizenga and colleagues (1987), at the Perkin-Elmer Corporation, stress the need for cooperation and communication in product design and put forward six steps:

- Both internal and external customers and their needs must be identified.
- Quality characteristics require identification.
- Means to measure quality characteristics must be obtained.
- Quality goals that satisfy customers and suppliers at reasonable cost need establishing.
- The process to attain the goals must be in place.
- Processing capability requires verification.

Six principles, according to Pearce (1987), are involved in a design assurance policy for food products:

- Design work for quality must conform to marketing concepts and regulatory needs.
- Design work for quality should conform to properly established procedures and standards.
- All design work should be properly documented for quality, with changes both recorded and regulated.
- Challenges of the design should be carried out at each stage of scale-up, including production trials.
- Third party review of design work for quality is required at critical stages before advancing to subsequent stages.
- A feedback system must be prepared to collate all activities and planning that support manufacture of new products.

Pearce's principles can be seen as an elaboration of Mayo's four points and Huizenga and co-workers' six steps. Pearce (1987) further breaks these principles into 12 subsystem requirements and develops a responsibility matrix for product design assurance, with primary responsibility delineated.

Wilhelmi (1988) considers the following requirements to be important in product design:

- composition of the product,
- safety considerations for the product,
- regulatory compliance,
- knowledge of product stability,
- packaging considerations,
- considerations in the market.

All of the above authors have essentially said the same thing concerning designing for quality. That is, they emphasize formulation for quality, stability, conformance, and safety, based on known spoilage mechanisms and anticipated abuse, and protected by suitable packaging. Such design is a built-in safety net: it is error-prevention in new food product development. Technologists must incorporate safety and quality design into their products through the judicious use of ingredients, processing, and packaging technology.

Equally important are control systems to monitor safety with respect to hazards of public health significance, to maintain the desired quality characteristics, and to stabilize the product during processing stages. Monitoring systems should already exist in the plant's good manufacturing practices, QC/QA systems, and HACCP programs, which are in place for the company's other products. However, these existing systems for maintaining product integrity must not be relied upon to be adequate for new products. A reevaluation of all quality systems whenever a new product is introduced into a food plant is imperative. If a reevaluation is not done, new products may introduce unsuspected new hazards, with the result that product safety of all the company's products may be jeopardized, which could lead to product losses.

Quality in new food products, with respect to whatever quality attributes, does have costs. These costs (overheads) can be crudely distinguished (Table 6) as indirect costs and direct costs of quality. Indirect costs are hidden in work to design new products to minimize hazards and to design processes to manufacture products with these designed features. Modifications to plant equipment may involve much work. Training of operators and technicians in the new procedures that may be required by new processes (sanitation, preventive maintenance, and process control) and new products (storage and quality control) contributes to indirect costs. Staff must be trained in new inspection routines to recognize and report hazards associated with something novel for them.

The direct costs include salaries of inspectors and analysts for grading, collection, and interpretation of data for incoming materials, in-line, on-line, and off-line process controls, and the costs of rework of failed material (that does not meet standards), returned goods, or lost customers. The more time, effort, and money that is spent on prevention in the design phase of new product development, the less the costs will be for inspection and removal of products not measuring up to standards (Table 6).

**TABLE 6. The Costs of Quality in New
Food Product Development**

Indirect Costs	Direct Costs
Quality design of product	Increased inspection
Product standards	Additional analysis
Ingredient standards	Increased costs for
Process design	Maintenance
HACCP programs	Sanitation
SPC programs	Hygiene
Operator training	Warehousing
Maintenance	Control systems
Sanitation	In-line
Hygiene	On-line
Equipment operation	Off-line
QC inspectors	Rework
Analysts	

Every food company has some policy respecting QC/QA. In most large companies, this may be in the form of a clear statement of the company's policy respecting the quality of its products. From this statement are derived procedures for quality in purchasing, processing, warehousing, and indeed, for every aspect of the company's business. The end result is a manual of operations documenting all the product and process procedures respecting quality. Implementation of these procedures is the responsibility of the QC/QA department.

Supporting a plant's QC/QA systems and HACCP program are several interrelated programs complementing the company's quest for safety, quality, and stability. These are:

preventive maintenance
pest control
plant sanitation
statistical quality control procedures
worker-related programs
  worker safety and health (ergonomics)
  hygiene for food handlers
  training programs for food handlers
grounds maintenance program
good manufacturing practices
environmental safety
  waste management
  water reclamation and effluent control
  odor reduction.

Manufacture of new food products in any food plant introduces new hazards into that plant. New ingredients and raw materials may introduce new and unknown microflora into the plant; larger quantities of known ingredients or raw materials may introduce pests that swamp the company's existing control systems. New ingredients may be introduced into a plant where the storage facilities are not adequate for them. Plant personnel unfamiliar with new ingredients, with strange raw materials, with unusual products, or even with new plant routines can introduce hazards. Purchasing departments may not have the expertise to purchase, for example, on the commodity market, or the plant may not have the facilities to store the ingredients properly.

An example can illustrate this point. One company introduced a flavored breading. The product was excellent. Complaints soon began to come in from consumers; rapidly the problem became apparent in the plant. A moth infestation had developed and spread into other parts of the plant and, more damagingly, into other products. The cause was simply explained. First, the purchasing department had no experience in purchasing bread crumbs and bought long. That is, they bought several months' supply and stored these in the plant. The storage area was a porous cinder block area — not the recommended building material — specially constructed for the purpose. The plant itself was not experienced in the storage, processing, and cleaning operations of a dry product.

Anything new or unusual introduced into a controlled environment like a food plant can contribute a potential hazard. The existing control systems in the above example were not designed for the novelty and the hazards associated with it. If the nature of the hazard is not understood or anticipated, then the hazard cannot be controlled or minimized.

A. Hazard Analysis Critical Control Point (HACCP) Programs

The early history of the development of the concept of hazard analysis critical control point system is described by Bauman (1990).

A HACCP program should be prepared for every food product that a plant produces. Before any new product is introduced into a plant for production, a HACCP program must be laid out by the developers. Even if the new product is merely a line extension of already established products, HACCP programs for these established products may not be adequate for the new one. HACCP programs are, or should be, flexible and able to evolve continually to meet equally flexible hazards of economic and public health significance. Archer (1990) cites the emerging microbial hazards, such as *Campylobacter jejuni* and *Listeria monocytogenes*, which can challenge an inflexible HACCP concept.

Development teams should work closely with their production members to prepare a flow chart of the process and equipment to be used for the new products. It is here that feedback from technologists can help production staff assess the effects that the rigors of the plant processes might have on the product and its sensitive characteristics. The goal is to identify hazards in the process that may jeopardize the new product's manufacture. They identify points in the process where, if control were to be lost, a hazard might arise.

Tighter vigilance at this point in the process must be exercised. It is here that the experience of production people, with their knowledge of the past performance of the process lines, can greatly assist the other members of the development team. They will know the limits of process equipment; they can more readily identify hazards.

Hazards can be either eliminated or minimized at that critical point in the process (or in the plant environment), or the product can be protected from the hazard at that point. Combined teams of technologists, production staff, and QC/QA personnel can now elaborate upon carrying out this control for elimination or reduction of identified hazards. That is, control limits for the guidance of production can be established with confidence.

The final step is the determination of which process parameters or which product-in-process characteristics to monitor for this control. A sound statistical process control program will then warn of deviations from accepted norms. Stevenson (1990) elucidates the steps and considerations that should be applied when introducing HACCP principles to foods.

B. Standards

Development teams have the responsibility to establish standards and specifications for ingredients used in new products, for packaging materials used in protecting the safety of new products, for the labels, label statements, and the placement of the labels on packages, and for a detailed description of the entire process, with the various pieces of equipment used and the support systems for the new products, including everything from the analytical procedures to be used to the detergents and sanitizers necessary to clean equipment of soil.

Specifications for the ingredients need to be clearly defined. First and foremost, purchasing departments need precise descriptions that permit them to obtain exactly the style, grade, cut, color, flavor, heat level, or whatever other criteria that are required at the lowest cost or at a cost consistent with the original cost estimate for the product. For this reason, product development teams must define as closely as possible all the important, essential characterisics of ingredients used in new products.

A word of caution is necessary here. It is wise to avoid using a supplier's specification sheet for an ingredient when writing these standards. Often, simply because a supplier's ingredient happened to be in the test kitchen at the time of development, that particular brand of ingredient eventually is written into the formulation and its specification becomes the standard. This practice limits the purchasing department to one supplier. If the price should increase or the supplier should be strikebound, production can be jeopardized.

Therefore, only the essential characteristics of an ingredient in a new product formulation should be identified. No two supplier's specifications, even for the same product, are identical. Thus, the wholesale use of a specification sheet of a particular supplier's ingredient may result in costly and unnecessary overspecification if a second supplier of a similar ingredient must be called in to supply to the original supplier's specifications. In addition, later generations of developers will not know what the important and necessary attributes of that ingredient were if these are identified only by a supplier's specification sheet. Other suppliers of that ingredient may not be able to meet a rival's specifications except as a special, and costly, production run. This precaution by no means obviates the need for recommended suppliers. It does mean that developers should not specify ingredients as "similar to the XYZ company's product": that is not adequate specification. If color is important, it should be so specified and the color identified; if particle size is important, the range of particle size should be identified.

After the important attributes (and only the essential attributes) of each ingredient are identified and specified, then a list of suppliers whose products meet these specifications as defined by the product development team can be prepared for the purchasing department.

A second important requirement of specifications is that analytical or functional tests be developed that permit the QC laboratory to measure or assess that ingredients, when received, meet the attribute standards required. Included with this methodology there should be sampling procedures suitable for the nature and form of needed ingredients. If attributes are specified for any ingredient or raw material, they must be checked frequently enough for safety's sake and frequently enough to keep suppliers honest.

Where applicable for ingredients and more particularly for raw perishable materials, proper storage conditions need to be specified.

There should, of course, be finished product specifications. These are not the same as quality standards. Quality was designed into the product. These finished product specifications identify whether the product has met the specifications required in the trade of that product.

C. In-Process Specifications

In a previous section, reference was made to the sensitivity of food products to treatments (such as various methods of heating or size reduction) to which they are subjected. These treatments can have a profound influence, for the better or the worse, on the product's quality characteristics.

To assure a consistent high quality in any food product, especially for new products, all treatments that can have an effect on quality must have tolerance limits specified. Since this affects virtually every unit operation in the plant, it is wise to document a product flow, identifying each and every piece of equipment used. Most important are:

- temperatures of the product at each stage in this product flow (heat damage to sensitive ingredients);

- times at these temperatures (flow rates, product viscosity, flavor losses, etc.);
- shearing action on the product (flow rate, pumping and mixer action and speeds, pipe diameters, energy consumption of motors, etc.);
- pressure changes on the product.

Any operation within a process that might have an effect on a product's character must be provided with operating limits.

D. International Standards

Ever-expanding markets are a goal for every company introducing a new product. Ultimately, this will mean the exportation of their product. There are no specific international standards for foods per se, but two bodies, Codex Alimentarius and the International Standards Organization (ISO), do provide guidelines for contracting companies, i.e., between manufacturers of products and buyers.

Codex Alimentarius publishes such documents as *General Principles of Food Hygiene* and *Recommended International Code of Hygienic Practices for Canned Fruit and Vegetable Products*. It would be advisable for any company introducing new products to be guided by the general principles outlined in these documents. Walston (1992), with reference to Codex Alimentarius, discusses many problems in international food trade. In the Uruguay Round of talks for the General Agreement on Tariffs and Trade, the suggestion that Codex be the standard for food safety has sparked controversy. Problems arise because of perceived shortcomings in Codex. There are three criteria for rejection of food in Codex: quality, safety, and efficacy. However, many governments, especially those of Muslim countries, and many nongovernment organizations believe that there should be cause for rejection on religious or ethical grounds. Such nonquality or nondefect related rejection of product could have serious implications for companies developing products for exportation. The product developer with an eye to exploring export markets would be well advised to be aware of international standards and regulations.

The ISO documents, ISO 9000 to 9004, are much headier material. ISO 9000, Quality Management and Quality Assurance Standards — Guidelines for Selection and Use, purports "to clarify the distinctions and interrelationships among the principal quality concepts ... and to provide guidelines for the selection and use of a series of International Standards on quality systems that can be used for internal quality management purposes and for external quality assurance purposes ...". ISO 9001 describes requirements for a quality system in which a contract between two companies requires proof of the capability of the supplier to produce to the required level of quality.

ISO 9002 lays out the requirements whereby, in a contractual arrangement, the supplier must demonstrate the ability to control the process within specifications. ISO 9003 describes the necessary quality system requirements for end product inspection and detection. ISO 9004 is really a guideline for establishing total quality management (TQM) in a company. It provides the bases for establishing and maintaining a quality management system. TQM is described by Shapton and Shapton (1991) and by Taylor and Leith (1991). The latter reference is accompanied by descriptions of the application of TQM at several food plants (pages 21 to 26 of the article).

Direct application of either the Codex Alimentarius or the ISO series of documents may not be pertinent in the development of added-value products for domestic consumption. Where exporting is a major objective or where companies wish to contract out the manufacture of their products to co-packers domestically or in other countries, the ISO documents, in particular, may be very useful in negotiations. Many companies now require that their suppliers and co-packers be ISO-certified. Possession of such certification may provide a powerful marketing edge. Ingredient manufacturers should be aware, particularly, of the ISO documents and their possible impact on sales.

IV. SIMPLIFYING THE WORK: USING COMPUTERS

Access to computers and associated software programs is well within the reach of large or small companies. Computerization usually starts in companies in the accounts department but eventually does find its way into technical departments and progresses, at last, into new product development arenas. The ready availability of software capable of performing a myriad of tasks permits small companies the same computational ability and power as larger companies. O'Donnell (1991) values computers in research and development programs of the food industry in costing of ingredients in formulations, nutritional computations for formulations, and experimental design to reduce the amount of experimentation. Developers, equipped with their computers and appropriate software, are now in a position to formulate a product to have a given nutritional value, to meet a nutritional standard, and to reach these goals with the least number of experimental tests.

Computers (and the software to use with them) serve three useful basic functions in new product development:

- information storage, management, and retrieval, which, if coupled with external data bases, provide a strong technical information base as well as a market information base to assist product design;
- a "number crunching" capability that allows the use of sophisticated statistical analytical programs to simplify experimental testing, sensory and consumer analyses, and least cost formulation techniques, to mention only a few;
- graphics programs that permit manipulation of three-dimensional solids, such as graphs, to complement statistical studies or to assist package and packing designs or to create and design food labels.

These functions or variants of them provide developers with tools to reduce the work load, to manipulate data for more efficient information retrieval from the data, to retrieve and classify data to see trends more clearly, to develop expert systems, and to communicate data and information more rapidly and efficiently.

A. Information Management

Without access to information, businesses would be helpless. Information about markets, legislation concerning trade, financial analyses, consumerism, and consumer trends provides many businesses with guidelines for strategic planning. Equally valuable for the food technologists is access to technical information. However, most food manufacturers do not have a staffed on-site library (Goldman, 1983). One way around the problem of lack of library facilities is to subscribe to a database. This can be accomplished either with the database on a disk or available on-line via a modem.

Williams (1985) provides a general discussion of the types of databases available: bibliographic databases, full-text databases, and numeric databases. As well, she describes and discusses aids to on-line retrieval, such as user-friendly front ends, intermediary systems that help the searcher with questions, and gateway systems that assist the searcher to reach other databases within the system. Access to technical literature through these databases reduces the time spent in literature reviews, provides a wide access to literature beyond the budget means of most food companies, and provides scientists in development with the technical literature support they need.

Buxton (1991) describes the equipment needed to go on-line with databases and the terminology used (in user-friendly language) in going on-line. Some general databases available in the U.K. are described.

Hill (1991) describes International Food Information Service (IFIS), which is a database producer. Its main product is the well known Food *Science and Technology Abstracts (FSTA)©*, which is available in printed form, on-line, on magnetic tape, and on CD-ROM. Its database contains original abstracts, authors' summaries, or, in approximately 10% of the cases, title-only entries. Classification in *FSTA* is tabulated, and on-line hosts for accessing the database are described.

A different database compiled for retrospective searching is described by Mundy (1991). Citation indexes are developed by the Institute for Scientific Information (ISI) based in the U.S. Well known to most food technologists is the *Science Citation Index (SCI)©*. This index comprises four inter-related indexes. *SCI* can be used to find recent papers on a particular topic that use earlier papers in that field; the subject index uses key words or phrases to provide a list of authors using these reference phrases in their titles. The corporate index provides information on what has been published by researchers at a particular company, which leads to an index of authors from that location. *SCI* is available on CD, on magnetic tape, and on line. ISI also produces the popular *Current Contents©*, which is available on line and on magnetic tape.

Leatherhead Food Research Association (U.K.), through their Information Group has produced *Foodline©*, which comprises scientific and technical, marketing, and legislation databases (Kernon, 1991). These can be accessed on-line, worldwide. The scientific and technology database with the acronym *FROSTI©* (Food Research on Scientific and Technical Information) dates from 1974. All entries contain a concise abstract of the article, with key word cross-referencing. The marketing database (*FOMAD©* for Food Market Data) is supported by two databases, *FLAIRS UK©* (Food Launch Awareness in the Retail Sector) and *FLAIRS NOVEL©*. The latter contains information about products described as novel with respect to some attribute and introduced anywhere in the world. *FOREGE©* (Food Regulation Enquiries), legislation data bank, provides information concerning permitted food additives world wide (Kernon, 1991).

Predictive techniques in microbiology were discussed earlier (Buchanan, 1993; Gibbs and Williams, 1990; Walker and Jones, 1992; Williams et al., 1992), and the value of computers for storing and sorting data was emphasized. Cole (1991) reviews databases in microbiology and their uses in predictive microbiology and, as well, discusses laboratory information management systems.

O'Brien (1991) provides a good overview of both food and food-related databases that are available. For readers interested in the topic or perhaps interested in subscribing to a service, some difficulties and problems associated with food databases are discussed by Pennington and Butrum (1991). They describe translation problems of food nomenclature and descriptions as well as taxonomy of foods from country to country. These are major problems in trade between countries. Fish and fish products have a rich variety of local, regional, and taxonomic names which present difficulties in translation. Klensin (1991) describes the use multimedia (and hypermedia) techniques to answer some of these difficulties described by Pennington and Butrum (1991). Photographs and illustrations could very well be transmitted by computers to assist descriptions (Klensin, 1991).

Stored technical information is useful only if it can be accessed. But if that information could be so organized and structured so as to be available to assist the user, just as the apprentice has the expert to lean on, it would be a valuable tool. Anderson and co-workers (1985) describe developments in intelligent computer-assisted instruction that are "... programs that simulate understanding of the domain they teach ..." and can interact with the student according to a strategy. This is, in essence, an expert system.

The use of expert systems in the food industry is comparatively new. An oft-cited example is the Campbell Soup Company's retrieval of the expert knowledge of their hydrostatic cooker operator prior to his loss to the company through retirement and the development of this information into an expert system (Whitney, 1989). The chief value of expert systems is in improved training programs for personnel, in having an "expert" on site in times of crises to prevent the introduction of hazards into the food, in the design of new processes, in developing better control systems for safer food, and in the preservation of the accumulated wisdom of experts. Bush (1989), McLellan (1989), and Herrod (1989) describe in detail what is involved in the development of expert systems and the immense amount of effort required in their production.

A less esoteric use of computerized information management is described by Cooper (1990). He describes the planning and implementation of a program to organize consumer complaints. The program could provide weekly summaries of all complaints, complaints broken down to product or brands, and complaints broken down to specific factories. Summaries of complaints are sent to factory managers and quality control managers as well as to the board of directors. Such information, always useful in maintaining good consumer relations and alerting the quality control department to a potential breakdown in the system, can also serve as a guide to either product improvement or new product ideas.

B. Number Crunching

Statistical software packages are available that have been a boon to developers in the manipulation of numerical data in diverse fields, from consumer preference studies to obtaining response surfaces from formulation trials. Software programs permit rapid analysis of multivariables to extract efficiently all the results buried in the data.

The analysis of sensory data has been effectively accomplished with the aid of appropriate software. A typical use of computers for assistance in the collection and analysis of sensory data is described by McLellan and co-workers (1987). They combined a computer system with software for sensory analysis, an optical card reader, and cards specially designed for collecting data from triangle difference tests, rank analysis, category scaling, hedonic scaling, paired preference/difference tests, and magnitude estimation. They estimated that in about 10 minutes they could enter a day's entire set of data from a sensory laboratory running four panels a day, each panel consisting of 30 judges doing category scaling assessments.

Thomson (1989) describes software for generalized Procrustes analysis (*Procrustes PC©*) for sensory methods relying on consumers rather than on trained panelists. A program identified by the acronym, *REST©* (Repertory Elicitation with Statistical Treatment) is also described, with an explanatory example, as a structured method of qualitative market research based partly on the repertory grid method and generalized Procrustes analysis. A more detailed description of the software used in these studies with these same two techniques can be found in Scriven and co-workers (1989), who applied the technique to study the context (time, manner, place, or circumstances) under which consumers drank a variety of alcoholic beverages. Gains and Thomson (1990) used the same techniques, generalized Procrustes analysis with the repertory grid method, to study under what contexts a group of consumers used a range of canned lagers.

Such data is invaluable in defining market niches for products and opening up new market opportunities. GAP analysis is a crude and unsophisticated variant of the above.

By far the greatest use of computers is associated with statistically based experimental design software. These software programs allow developers to take calculated shortcuts in the number of trials dictated by classical statistical experimental design. For example, using a factorial design, the number of experiments mushrooms rapidly as the number of variables and the levels at which each variable is to be tested increase. Thus, the number of trials required in a factorial design where v is the number of variables and L is the number of levels at which each variable is to be tested is L^v. Astronomical and costly numbers of trials are required to test four ingredients of a product formulation at three concentrations.

Mullen and Ennis (1979) describe the design for applying a linear equation process to a computer program to produce a six ingredient hypothetical product, which supplied 10% of its calories from protein, 35% from fat (high by today's standard), and 55% from carbohydrate. Mullen and Ennis (1985) later refined their program (which was available from the authors) to handle 15 variables but reduced the amount of experimentation by using fractional replication. The procedure is described in detail.

The history of experimental design, as well as an explanation of the principles of experimental design as an aid to product development, are described by Dziezak (1990). She describes screening and optimization designs and lists available software to accomplish these techniques for product development.

Optimization designs are particularly useful, as these permit the developer to investigate the optimum levels of ingredients to maximize a particular quality feature or to alter a process to get a maximum effect (or a minimum effect if the effect is undesired). Two techniques are used: response surface methodology designs and mixture designs. Henika (1972) used the example of improving the wettability and flavor of an instant breakfast cereal product to compare and explain classical testing vs. the response surface methodology approach in getting answers more quickly and cost-effectively. Hsieh and co-workers (1980) developed a synthetic meat flavor using response surface methodology. Mixture designs treat the unique problem of formulations whose proportions of all the ingredients must equal 100%. Designs for optimization using response surface methodology and techniques based on mixture designs are available from software suppliers (Dziezak, 1990).

Further examples of response surface methodology in the optimization of processes can be found in work by Bastos and co-workers (1991) who upgraded offal processing by extrusion cooking to produce a finished product with good solubility and emulsifying capacity. King and Zall (1992) used a model system to study low temperature vacuum drying using a three variable, three level design.

Skinner and Debling (1969) describe the application of linear programming techniques to management decisions in food manufacturing. They explain three problems in food manufacturing, with examples of each worked through:

- the allocation problem faced when several products use the same commodities, which are available only in limited supply in their formulation. Which product to make? Their example concerns fruit salad vs. fruit cocktail;
- the blending problem that arises when ingredients of a particular product can be blended either to meet a quality standard or to achieve a cost standard (applicable to least cost formulations). Which proportions to use? A sausage formulation is used;
- a simultaneous blending and allocation problem, exemplified with a pork vs. beef sausage example.

The examples are worked through, but, as the authors state, a statistical software package with linear programming could be used. The blending problem and variations of it are ones frequently encountered in new product development.

C. Graphics

Computer graphics capability has grown immensely, as anyone interested in the subject of virtual reality can attest. Architects now plan a building in their computer with suitable software. Then, figuratively, they can stroll through the building, noting the views from different aspects from within the simulated structure.

Dziezak (1990) demonstrates some of the three-dimensional and contour plots that can be accomplished. Using computer-generated graphics to investigate response surfaces permits a rapid assessment of the avenues of investigation that would appear to be most rewarding. Rotation of these solids can reveal more data.

Bishop and co-workers (1981) make the case for the value of three-dimensional graphics in food science applications. The programs they developed and used in their applications are described and discussed. Floros and Chinnan (1988), using optimization techniques based on response surface methodology, demonstrated the application of graphical optimization to a process for the alkali peeling of tomatoes. By their technique, they were able to improve the quality of the product by studying response surface plots to select the optimum concentration of alkali and temperature.

Roberts (1990) demonstrated the use of computer generated 3-D graphics in predictive microbial growth modeling. While predictions may not be precise, he claimed that knowing the trend of growth would be highly important in designing stabilization systems for new products.

An obvious application of computer graphic techniques is in the design of packages and labels. Lingle (1991) describes the use by several food companies of computer-based systems to control package and label design. The advantages cited are:

- ability to manipulate designs in any desired fashion for application to line extensions or to redesign packages;
- ability to store designs more easily on disk rather than as paper art;
- ease with which images can be communicated with others for decision making.

Graphics software is currently available that allows a company to bring package design in-house.

Chapter 6

The Final Screening:
Going to Market

To market, to market, to buy a fat pig,
Home again, home again, jiggety jig.
To market, to market, to buy a fat hog,
Home again, home again, jiggety jog.

As readers have learned, techniques for consumer research vary from small focus groups being surveyed for opinions on products up to large assemblages of consumers testing products in their homes. It becomes a moot point as to where focus groups end and personal interviews conducted in church halls or mass surveys in rented cinemas should begin. There is a progression with just-noticeable-differences from focus groups to large-scale consumer surveys. Likewise, when product is being sampled, where does the so-called minitest market begin and end? And where does the traditional test market begin? What's in a name, whether the test is carried out in the homes of two or three hundred consumers, in a single food retail chain, in several stores in each of one or two cities, in stores in two or three counties, or in stores in a region of a country?

Binkerd (1975) describes the variety of tests that would have been applied to the product so far as the following:

* focus group interviews
* concept screenings
* blind product tests
* concept tests
* minimarket tests
* test markets

Whatever nomenclature is used for these tests — and readers will find great variation in the literature — companies have progressed through several stages of testing with consumers. First came concept testing, which may have been accompanied by a laboratory sample of the product in focus groups. Then, the next contact was with product testing by selected groups of consumers. These might be central location tests or in-home use tests. Frequently, products being tested at this stage were not factory-run products. They were superior products prepared "differently" in pilot plant or test kitchen facilities. There may have been further testing after this, in what some companies call extended-use tests or minitests. These are small, highly controlled market tests conducted in well-defined locations, which can last a few weeks to several months.

Test markets will be the first opportunity, in a controlled manner, to see how consumers will react to new products. They are tests: they are still a phase in the development sequence. In this phase, development teams will put their products into competition with other products. How will consumers react to the product? What retaliation can development teams expect from the competition? How controlled will the test be as a result of the competitions' retaliatory action? Product development now enters the arena of the marketplace, and the game becomes tougher and deadlier.

At this point, marketing staff have a clear picture of their targeted consumer. Advertising and promotional strategies based on consumer research, either conducted in-house or contracted for with outside companies, are in hand. The production department has filled the pipelines (distribution channels), and the timing is right for a market launch into a test market. The ball, now, so to speak, is in the marketing department's court. The next stage in development is theirs.

Companies have committed major amounts of money to advertising and promotional materials, to new equipment, and to facilities with which to make the product (see Chapter 1, Figure 3b and Chapter 3, Figure 8). There is a very high cost to pay for a test market. Everything pertaining to the new product is to be tested: the research department's skills in stabilizing the product, the production department's ability to manufacture a uniformly high quality product, the shipping department's ability to distribute the product in top quality, the marketing department's skills with its direct advertising and promotional campaigns, and management's tactical skills at countering competitive action.

I. TEST MARKET

Test markets are launches (introductions) of new products into regions carefully selected for a variety of geographical, marketing, and company reasons. After a predetermined period of time for introduction and repeat sales, results of the test are analyzed. On the basis of these results, marketing is either continued for further data collection, marketing is extended into other regions, or the product launch is dropped and the whole project reevaluated.

Two examples suffice to illustrate the many differences in the conducting of test markets that exist. Mazza (1979) test marketed native fruit jellies in a gourmet gift pack through one retailer's stores in two cities in the first year of introduction. In the second year of introduction of this seasonal product, three retail outlets were chosen, and two additional cities were included. Marketing and consumer evaluations were carried out through a questionnaire accompanying each gift pack sold in both introductions. The questionnaires were to be completed and returned. Response averaged approximately 20%, and from the data, information was obtained on why the product was purchased, what attracted the purchaser to the product, whether the purchaser would repeat purchase the product, and how the purchaser would rate the product.

Clausi (1974) describes how the then General Foods Corporation moves to test market. After modifications arising from the results of in-home testing have been made to a product, the product is put into test market in one or two cities. Then, depending on market data, a move to a larger section of the country is made to evaluate both consumer reaction to the product and awareness of the advertising and promotional campaigns.

Test markets, quite simply, are large-scale commercial marketing experiments in a geographic area of the country chosen for very specific market-related reasons. Test markets are still part of the screening process for new products. They provide their own unique opportunities for further experimentation with the product, its package, and its message put forward in advertising. It is a very complex experiment involving all the following people and reasons:

- food technologists, who want to verify (and to see justified) all technological skills used to establish the quality and stability of the product and to determine the reliability of the package vis-à-vis all the abuse that products in market channels can encounter;
- production staff, who want to see how well they can maintain quality and product standards to meet consumer demand under full production conditions;
- marketing staff, who want to evaluate advertising, package design, and promotional materials for effectiveness, to evaluate different advertising media, to gauge consumer reaction to the product for reevaluating the product's positioning, and to estimate sales potential to determine the viability of the project for the financial department;
- management-level personnel who want to evaluate the sales force's efforts and to note the retaliatory action of the competition in the marketplace;
- traffic department staff, who want to evaluate the distribution and delivery systems for keeping shelves stocked and deliveries on time;
- financial department personnel, who will be constantly examining the price structure at the manufacturer's, trade, and retail levels and projected profitability.

Consumer research is as active during test markets as it was during earlier stages of development. Consumer reaction to the product — how and when the consumer uses the product — are questions to be answered. Is the consumer misusing the product? Is the product's message being misinterpreted? Are preparation instructions clear? Consumers' reactions to and usage of products may point to new opportunities for repositioning products. Test markets are learning processes.

The nature of test markets varies widely, depending upon the type of product to be tested and the goals of companies doing the testing. (Types of new products and their characteristics were discussed in Chapter 1.)

Improved (reformulated) or repackaged (established) products, for which new market niches are being explored or for which new marketing strategies are being tested, present unique test marketing situations to marketing personnel. These products are already established products, so to speak. In test markets, marketing departments will be seeking comparative data for their consumer research: will changes incorporated in "established" products be accepted by established consumers? Will they attract new consumers? Will new market niches be opened?

Introduction of line extensions or "me-too" products into test markets can cause some concern to marketing personnel. There is very little to make a claim about with such new products in test market. Heavy advertising on line extensions may cannibalize the company's existing products. If already established products are valuable cash cows to companies, then marketing personnel must interpret consumer reaction carefully, not only for the new product but for the remainder of the product line. Pushing "me-too" products to consumers gives impetus to other, competitive, "me-too" products. These test markets must be designed and analyzed to clarify what activity is going on in the marketplace.

Test market situations for small companies and large companies can be very different. Development teams of large companies are much more conservative, cautious, and afraid of making a mistake. This conservatism permeates collectively throughout large companies (more is at stake should a blooper be made) and individually down the chain of development command from vice presidents of research and development to product managers. People may have something to lose: their jobs.

Small companies can introduce their products at county fairs or local sporting events with tasting sessions. McWatters et al. (1990) used a mobile kitchen to evaluate consumer response to akara, a snack product, in such minitest markets. Small companies using this technique get immediate feedback of consumer reaction to their products' characteristics. Their flexibility allows a rapid response to consumers' reactions by reformulation, repackaging, or refinishing. They can literally deliver and sell off the back of their company station wagons to small independent grocers or independent franchisees. They have the patience, the time, and the proximity to develop a market, something the big company does not have.

Some companies challenge the value of a test market, although, in truth, much of the challenge may center around how companies define test markets.

Many companies accomplish their new product marketing research with minimarket tests with a much cheaper outlay of money than a full scale market test.

Actually omitting a test market is a very daring gamble, but in certain circumstances it may be justified. Where it is essential to get a step into the market before the competition, to be on the shelves first, eliminating the test market may be right. Such a marketing move could provide up to 6 or more months of lead time in which to build a dominant market share. This is risky but perhaps worth the challenge in a sound, economically healthy, flexible company. An obvious disadvantage is that the company has no opportunity to evaluate its advertising and promotional campaigns.

There are many concerns in a test market, and these will be now considered. The where, when, and how to introduce are all factors that are closely interwoven. Separating these in this manner should be recognized for what it is, an explanatory device. The nature of the product being introduced will greatly influence the answers to these questions.

A. Where to Introduce

The *where* of an introductory test market could be the subject of a book in itself. Marketing personnel want unbiased marketing data from which to derive information for making very important, economic decisions. As much as possible, the data they obtain should not introduce biases stemming from the location of the test market.

First, is the area chosen for the test market peculiar to the company? And, equally important, is it peculiar to a competitor? Answers to both questions are vital considerations for the interpretation of sales data. Large companies are much more concerned with these answers than are small companies. For example, introducing a product into areas where a large company is well known, such as a factory town, or into areas where the company has not been a good corporate citizen — where the company has been a local polluter — may lead to distorted sales and marketing data. Product sales may be influenced by the large company's reputation. Hence, skewed information on which to base sales projections to other regions of the country will result.

Small companies rarely have any geographic area that is peculiar to them. They test market within range of their factories for fear of stretching their distribution resources to the limit. Usually too, they avoid relying on other sales forces than their own. Their size usually precludes them from being poor corporate citizens. Their tactics in test markets dominated by a single competitor could go unchallenged or even unnoticed.

Second, introducing a new product in marketing areas that are heavily saturated by a major competitor's products is foolish unless, of course, that head-to-head confrontation with the competitor is deliberate; however, this is not usual practice. A test market is neither the time nor the place to challenge competitors; it is an experiment, an expensive experiment. As such, a company has established an experimental design, a marketing plan, to obtain as much marketing, sales, and consumer research data as possible. The plan must be adhered to by the company. But if the launch challenges competitors, they can be expected to retaliate in some manner that will seriously bias the test market results. The competition should not be expected to follow the company's marketing plan. Indeed, one can expect the competitor to disrupt the test market with its own tactics.

In addition, a market dominated by a competitor will be costly to penetrate. Heavy advertising, promotions, and trade allowances to obtain any shelf space or significant market penetration will be a burden on profits.

Third, the area chosen for introduction of new products should be one where there is a good sales force in position and a good distribution system already established, especially if the product has a short shelf life. The sales force should neither be a strong sales force nor a weak one. It is not the strength of the sales force that is being tested.

Fourth, evaluation of advertising, promotions, and sales efforts, as well as of the product itself, is an essential part of the test market and the new product development process. If the area chosen for the launch is dominated by large retailers or by a single large retail chain, it may not be possible to do this well in such an area. Dominance by large retailers or by single retailers restricts the activities of marketing personnel in planning promotions and advertising campaigns. Campaigns may not be conducted as a company may wish but as retailers want.

Small companies are at a distinct disadvantage in trying to introduce products where a large retailer dominates. The manner in which large retailers conduct their purchasing does not favor new introductions by unknown companies or one-product companies without heavy advertising support. For small companies, introductions are usually done with little fanfare in small independent grocery stores.

Fifth, test market areas should not be dominated by monoeconomies. That is, the area should have a mixed economy. The economic health of an area can have an influence on the buying habits of consumers. If the area is depressed, consumers may not be willing to try new products or may not want to purchase higher priced, added-value products that are perceived as being luxury items. On the other hand, consumers can be very perverse. Marketing departments should take cognizance of the economic mix of communities where new products are planned when interpreting market data.

Sixth, is the targeted consumer in the chosen test market area? Introduction of products with a strong ethnic appeal in areas devoid of that ethnic group would be a remarkably stupid marketing ploy. Likewise, introducing products aimed at an elderly population into a growing suburban area dominated by young families does not make much sense either.

Seventh and finally, is the product peculiar to the test area chosen? Distinct biases toward different flavors, colors, forms, or styles of products have very specific regional preferences. One need only consider the many different styles of "authentic" chili there are: meatless, "con carne" with ground meat or with chunk meat, and those with heat levels from mild to very hot. Pizza has many variations of styles of crust and shapes, for which there are regional preferences. Each area with its unique style considers theirs to be the authentic one. This variant becomes the criterion against which new products will be judged. Two popular regional Quebec dishes are quite distinct to the province: *poutine*, a mixture of french fries and cheese curds topped with gravy and *cretons*, a form of meat spread made of pork ("more fat than lean"), drippings, spices, and bread crumbs. These regional dishes, which are delicious, would find little acceptance outside Quebec.

B. When to Introduce

The seasonality of products dictates when test markets are carried out. Promoting a soup to be consumed hot in the summertime or promoting ice cream or frozen yogurt when the snow is flying outdoors are both examples of inappropriate timing for an introduction. But it is not weather alone that determines seasonality. Promoting seafood products during, for example, the American Thanksgiving period or over the Christmas period would be unwise marketing choices for the timing of introductions. Products associated with national festive occasions or ethnic festive occasions should be introduced at the appropriate time.

Snack products associated with activities such as the return to school and the need to pack school lunches should be introduced to coincide with that season. Leisure summer activities are associated with foods like salads, prepared meats, dips, and the like, designed for outdoor or patio living. Wintertime outdoor activities bring in an entirely different range of products. The timing for the test market, the food itself, and the activity associated with it must fit.

When to market is closely related to another question concerning time: how long should test markets be continued before an evaluation is made? The simple answer is that the test market should continue until reliable data has been obtained to evaluate sales volumes, the effectiveness of advertising and promotional strategies, and consumer response. Test markets must be long enough to measure the consumers' reaction to the product. This period must include sell-in to the trade, promotion, first purchase by consumers, and repeat purchases. Time is necessary to establish a pattern of purchasing, both by the trade and consumers. Further time is required to analyze data and transfer information back to marketing personnel and to the other members of the development team for any refinement of marketing strategy.

Yet, too long in limited test market without capitalizing on the advantages of early introduction serves no useful purpose. Lead time is lost. Copycat products may be introduced in other market areas by the competition, obtain a market share in these new areas, and make further market penetration and expansion difficult. Timing is very important.

Even if one has the best-laid plans, events can disrupt them. Events in international trade, agricultural pricing structures, or other events in the food industry can occur that produce a short-term alteration in consumption patterns or in the economics of an industry. Shortages can occur, with the resultant increase of raw material prices. For example, the launch of a precooked bacon product was seriously disrupted when the availability of pork bellies declined and prices rose (personal experience). Countries can place trade embargoes on goods or buy up stocks, causing temporary shortages and, consequently, rising prices; or they can flood the markets with commodities, driving prices down. Natural disasters can also play a short term economic role in raw material shortages. Such short-term events can greatly influence the timing of test markets and be very disruptive of tests already underway.

C. How to Introduce

How to introduce products can be a problem for both small companies and large companies. Small companies find it difficult to get shelf space, and sometimes it is difficult even to get an appointment to meet with the purchasing agent of large retail food chains.

Stores are becoming much more discriminating, hard-nosed some might say, regarding new product introductions. New products take away space from products with proven sales records. Stores want to eliminate slow-moving items with poor margins. They are, therefore, reluctant to take on new products unless they are assured of good margins, rapid inventory turnover, and advertising and promotional support of the new products. A catch-22 situation can result; products that stores want can only be developed by market testing. Manufacturers of new products need space in stores for market testing, which stores are reluctant to give.

New product introductions by large companies are accompanied by extensive advertising and promotions. There may be in-store demonstrations, couponing in magazines, newspapers, door-to-door fliers, piggy-backing offers, and special pricing offers. Small companies cannot afford this.

Advertising and promotional activities must be measured for their impact on the introductions of products in any marketing area. Introductory promotions to consumers (and to the trade) are one-time events. Heavy introductory promotions cannot be carried on throughout the product's life cycle.

To interpret the volume of sales of the initial introductory period, the marketing department must understand the impact of in-store promotions, demonstrations, and couponing activity on sales; it must be aware of the price differential vis-à-vis the competition and must know what the competition was or was not doing during this period.

D. What Product to Market?

First impressions are always important and can be very difficult to change. This is especially true of first impressions in the marketplace.

All too often, a company will introduce a product into the market in a specially designed package or even as a specially manufactured product. It is not the normal product that consumers will see in repeat sales (if there are any repeat sales). Use of this special product can be a disaster. Consumers have been introduced to, have become accustomed to, or have come to expect something specific. A new package or modified product constitutes another new product. Data obtained on the sales of the product originally introduced may not be valid for the changed final product.

The test market must be carried out with the same product as the factory will run. Pricing deals, couponing, and special packaging can be necessary facts of marketing introductions. However, these are not permanent features that will continue throughout the life cycle of the product. Therefore, their impact on consumers and retailers must be clearly understood. Consumers have been educated to a particular price, package, and product. Retailers have come to expect certain price deals and promotional support.

II. EVALUATING THE RESULTS

At some stage during the test market period, there must inexorably come a time of reckoning, of evaluation while marketing of the product continues.

The foregoing litany of caveats should indicate that many factors can influence the results of test markets. Data can very easily be misinterpreted. The natural desire to want to see the product succeed can color judgement. Interpretation is a task not to be taken lightly. Clausi (1971), detailing errors in interpreting marketing data during the introduction of a dry cereal with freeze-dried fruit, commented "... the strong initial purchase pattern coupled with overwhelming consumer acceptance of the concept tended to obscure the significance of the negative evidence." Repeat sales were flat. Negative signs were not read correctly for what they said.

Why is consumer research and test market data so easily misinterpreted? First, understanding all the forces at play in the marketplace is very difficult. There is the consumers' behavior to understand and the activity of the competition during the test market to determine. All data obtained must be read against the backdrop of this behavior. Even something as simple as knowing whether the competition was suffering through a strike or had a shutdown because of some catastrophic fire during the test market period can assist in the interpretation of sales data. After all, a product's success in a test market location may be due to the nonavailability of competitive products on the shelves as much as to the new product's desirability. Knowing what consumers are doing, as well as what the competition is doing, is vitally important.

Second, personal feelings and emotions can frequently blind one to the reality of test market data. Product managers, in particular, can become very attached to pet product development projects, which perhaps they or others overpromoted to superiors during earlier phases of development. Justification for past actions may be read into their interpretation of the introductory results. Emotions must be eliminated from the interpretation of data.

And finally, even with all the data in place, the science of consumer research is still not precise enough to prevent misinterpretation of the information obtained from that data.

Criteria for evaluating whether the launch was a success or failure must be established. Different criteria will be used according to the objectives of the company performing the test market launch.

Four measures can be used:

- payback: when will there be profits?
- sales volume: will sales volume goals or targeted percentage share of market or even significant market penetration be achieved?
- consumer reaction: did consumers like the product, and how can this be capitalized upon?
- tactics: being there is the thing.

The first two are very similar. One says it in money, the second in volume and share of market.

First, one faces up to the financial reality of the project. Will it meet the financial objectives of the company as measured by the expected return on investment projected from sales? Developmental costs plus the costs of production and test marketing must be compared to the anticipated rewards for all those costs. The financial criteria may include not only an expected return on investment but also a time limit to achieving that return! No longer is a return on investment within 3 years considered satisfactory. Indeed, the time frame when the return can be expected may be the more important criterion to indicate success or failure of the project. An unrealistic rate of return can be the downfall of many new product ventures. The interpretation of "unrealistic" rests with the "bean counters" in the company.

The second measure is, of course, the volume of units of the product sold and how repeat sales figured in these. If projections indicate the volume of units sold can be related to enthusiastic consumer acceptance and repeat sales, then appreciable economies of production can be anticipated by scaling up manufacturing. Usually, on the evidence of learning curves, more units of something can be made more economically than fewer units of the same thing (Malpas, 1977). This, in turn, will influence the rate of return on investment.

The volume of sales must be examined very carefully to determine precisely what it means. If these sales are consumer sales, this is a very positive factor in interpreting the results of the test market. If they are merely case movements between warehouses or buy-up by the competition, sales volume could be deceptive as a measure in interpreting the results. It is not unknown for the competition to buy product so that sales figures would be distorted. Such a somewhat similar situation happened to me. I was on an acquisition study of a company, test marketing a line of pouch-packed entree items. The product was being test marketed in three large supermarkets in a large city. Prior to my meeting with the principals, I purchased two cases each of the four flavors for shipment to my client's laboratory. Much to my surprise, during my meeting, the president of the company regaled me with the tale of eight cases of product being sold in one store, so great was the demand for the product!

The third measure is an assessment of consumer reaction to the product. The reason why the consumer is buying the product and how the consumer is using the product, that is, the context of product use, are valuable. Can the strengths of the product be capitalized upon? Are refinements needed or future line extensions justified? Are promotion and advertising directed properly? Should other market niches be explored? The test market should be treated for what it is: a commercial experiment in consumer studies. Weaknesses in the product can be ferreted out by consumer research and eliminated before new marketing strategies are undertaken.

The last measure is more in the nature of a business tactic, a strategic tactic forced on the company. A product may be introduced or positioned to counter the activity of a competitor or to establish a position in a particular niche. Success would be measured by whatever small share of market could be gained as a foothold for strategic or tactical purposes at a later date. The company feels that it cannot relinquish a position in the marketplace.

Small companies can be more flexible about which criteria are applied for success or failure. Varying proportions of the second and third measures are mostly applied; if each week more units are sold than the week before and the store owners say consumers like it, then that product is a success. Small companies are less concerned with market share, which large companies use to define success; how much market share was obtained is not a question small companies ask themselves. There can be more patience demonstrated by small companies and more fine tuning of the product as sales and development proceed hand-in-hand.

In small companies, too, budgets for research and development are frequently not separated out, as in large companies. These costs may be bundled with quality control or production expenses. Consequently, development costs for new products cannot, with accuracy, be determined. Presidents of small companies are quite content to pay their bills and have an increasing bit left over each week.

At the conclusion of any test market, data on both the product and consumers are gathered and analyzed. Questions related to the product itself, its protective package, the label, preparation or recipe instructions, pricing, and positioning of the product need to be asked. Salesmen must be interviewed; consumers must be surveyed about their perceptions of the product. The need for changes in the product, the process, or the package must be reviewed.

Whether the test market was a success or a failure and reasons for the success or failure need to be examined. If reasons for the success of a product can be crystallized out, perhaps they can be applied in the future to another product. If the product is a failure, the same canvassing of information is required. Now, the question is, "Why?" What went wrong? What elements of consumer research, technical development, and analysis of data led to the failure? Here, in this analysis of a failure, companies, and their product development teams in particular, are not looking for scapegoats. These exercises should not be witch hunts. Too often, unfortunately, this backward analysis turns into just that, a witch hunt, with everyone trying to obfuscate the facts, with the result that nothing is learned. The search should be for flawed systems or faulty information that led to incorrect decision-making.

III. FAILURES IN THE MARKETPLACE

Most new food products fail to survive their first year in the marketplace. This loss is staggering when considered in terms of the wasted time and effort of skilled personnel and the waste in terms of monies spent in consumer research, equipment, ingredients, and packaging material, not to mention monies spent on advertising and promotion. Careful consumer research, thorough collection of data, and unbiased, detached, and unemotional interpretation of the data, coupled with sound product development should have made failure most unlikely. Hindsight, unfortunately, provides better vision than foresight. Hindsight allows one to make generalizations (or speculations) about what went wrong.

A. Causes of Failure

It would be much more productive if this section could have been entitled *CAUSES OF SUCCESS*. An in-depth study of why a new product succeeded would have been a much more valuable contribution to an understanding of new food product development. One would then be equipped with guidelines to follow for future product development: a series of "if this, then that" conditions would simplify the process of development. Causes of failure merely provide developers with a series of "don't's". However, predicting the success or failure of any product against the volatility of the consumer in a changing marketplace is still an art. As Clausi (1971) might have said it; reading all the evidence correctly, including the negative evidence, will lead to success.

It is difficult to classify the reasons for a particular product's failure and put these easily and neatly into pigeonholes, as Clausi (1971) found. Clausi, in describing one particular failure, could only suggest that the signals from the marketplace were misinterpreted. An examination of specific product failures can provide a very broad overview of probable causes, from which only generalizations can be made. One cannot apply the generalizations at the start of the development process or at any other point up to and including the test market and say that this product or that product will fail because ....

Simplistically, the causes for failure can broadly be classified as those the company could not have done anything about and those they might have done something about. The former are reasons or causes beyond the control of the company and usually are external to it (nevertheless, the company should have been aware of them and been forewarned). The latter are causes usually found within the company. These internal causes for product failure are not always manageable for various company reasons.

Rarely can one isolate a single factor as the cause of a product's failure in test market. Many small problems, only partly assignable as either external or internal, can trigger the failure of a product. As well, separating reasons into categories, such as external and internal, cannot always be accomplished with clarity. For instance, too small a market (an external reason) can be a cause for a product's failure. That is a reason beyond the company's control. But if that were the case, then should not marketing personnel have seen there was too small a market? Were the marketing capabilities and resources within the company either incompetent or inadequate (internal reasons)? How else would marketing research have failed to determine the magnitude of the market beforehand?

Thus, if one states baldly that one reason is external, one must, equally, understand that an internal reason may have contributed directly or indirectly to it.

1. External Reasons for Product Failure

After products have been introduced, marketing personnel may find that markets for them are too small. Growth potentials would be limited; possibilities of recovering development costs would be minimal. In certain markets, this knowledge may come unexpectedly. For example, changes in the purchasing policies of governments with respect to institutional buying for the military, for government-run correctional institutions or prisons, or for school meal programs may suddenly and abruptly be altered, and the size of a market may change. Nevertheless, companies servicing such markets should keep themselves informed of pending government changes by close liaison with their government contacts.

The difficulty of establishing footholds in markets dominated by a single competitor may result in product failure. Consumer acceptance can be too costly if advertising and promotional dollars must counteract retaliatory action by the competitor. In addition, the competitor is in a position to control retailers in such a manner as to limit shelf exposure for rival products. Companies introducing new products may find themselves not battling for consumers but battling competitors.

Domination of markets by a single customer (for example, a major retailer or the government, i.e., military) can present severe challenges to companies introducing new products into those markets. The cooperation of customers (retailers) is always essential, but when suggestions from customers become directions, then the situation can be fraught with stumbling blocks. After all, companies, not customers, have spent the development dollars and will risk most in a failure. But, the dominant customer has the greatest say in pricing and marketing stategies in general. The tail, i.e., the customer, will wag the dog, the developer. Often, the food service industry is one where this problem can arise. Also, the immense buying power of some retail chains has permitted them to dictate to producers what products and development they want for their marketing purposes.

There are product-related reasons for failure in the marketplace. With the introduction of a "me-too" product into a market that is saturated with similar products, consumers will refuse to buy another brand or variation if they cannot see a point of difference between it and already established products. The problem is beyond the control of the developing company; no one could have foretold the flooding of the market with copycat products.

Where technical novelty has been designed into a product and this technology is the dominant message to the consumer throughout introduction, the consumer can be forgiven for questioning, "So what?" "What advantages are there for me?" New forms of a product, such as frozen for canned, tablets for powders, aerosols for liquids, and so on, may fail disastrously if the consumer cannot see the advantage being offered or if the advantage (point of difference) over other similar products is insignificant. Flavored ketchups have not been successful for this reason. However, where the market is unsaturated, the same product may have some hope of growth. Markets can fragment and provide a special marketing niche for new "me-too" products.

Products ahead of their time, products for which consumers are not prepared, have poor chances of market acceptance. They meet consumer resistance because consumers have not been adequately educated to their possibilities. Arguably, one might consider this an internal reason for a market failure, one within the control of the company. Marketing personnel did not promote the product correctly. On the other hand, educating consumers is costly, and consumers can be quite quixotic. Why educate consumers only to have a competitor reap the benefits?

2. Internal Reasons for Product Failure

It is too glib to state plainly and simply that the intrinsic reasons why a company fails to launch a new product successfully are bad management, more bad management, and a little admixture of bad communication. Failure can be said to occur because of the inability of management to recognize the strengths and weaknesses of the company. Consequently, management fails to understand what business the company is in. This, in turn, leads to a lack of clear company objectives with respect to growth and new product development.

Marketing and research and development resources suffer as a result of the absence of clear company objectives. These resources are not adequately developed. Obviously, inadequate marketing resources within companies can lead to new product failures because marketing departments are incapable of or incompetent at conducting reliable market and consumer research prior to committing any research and development efforts to the project. More damning, the company's marketing people may be unable to recognize their own shortcomings. This is unfortunate, since many independent market research companies are available that could assist.

Lack of production capacity can be a cause of product failure. If consumers cannot get product because the plant cannot produce enough to satisfy demand, then the impetus of the launch has been lost. Consumers will be disappointed. But again, one must question the lack of foresight in management not to have seen this likelihood and backed up production capacity with copackers.

The escalation of research and development costs in new product development is a common cause of product failure. (It is linked very closely with the external cause of inability to recoup these expenses.) It is, perhaps, more readily understood than the other causes because of the human element entangled in it. People, whether in small or large companies, can become emotionally involved in their project. It is their pet, their "baby". Reasoning goes something like this: too much money has been spent to date to stop the project now, so spend more money to rescue the project. It is like the gambler who gambles more money to recover his losses. Costs need to be evaluated regularly to prevent their escalation. (This will be discussed in greater depth in the chapter describing organizing for new product development.)

A product can fail for technical reasons. It simply does not perform as promised or does not live up to the standards promised. The cause for the poor performance could be inherent in the product itself. That is, it was poorly designed, or, one of its ingredients may have been incorrectly chosen for the task it was meant to do, or the packaging may have failed to give the proper protection.

It is an oversimplification to suggest that there are only two reasons for the failure of a new product. Nevertheless, there can be a great deal of truth in such a generalization. These two reasons are expecting too much and not being lucky. The first, expecting too much, does not happen to companies whose objectives are based on a realistic assessment of their companies' strengths and a realistic financial or marketing objective. And good luck, or whatever one wishes to call it, comes more regularly, rather than randomly, to companies that utilize their resources well in order to research markets, consumers, and their products. At the very least, those companies will reduce their margins of error.

Chapter 7

New Food Product Development in the Food Service Industry

Two areas in the food industry are sufficiently different to deserve separate treatments with respect to new product development. These are the food service industry and the food ingredient industry. Both present unique challenges at each stage of product development, and these challenges are worthy of some further study.

Companies manufacturing or supplying in both areas need to have new product development programs with clearly defined objectives and strategies for obtaining these objectives. That is, of course, a reality for all new food product development work.

In both the food service industry and the food ingredient industry, similar to the food retail industry, there are customers whose needs are to be met. But there all similarity respecting customers ends. In this text, an attempt has been made to distinguish between "customer" and "consumer". Very simply, a customer buys a product to use in another product for resale; a consumer buys a product for ultimate consumption, that is, for eating.

Who, then, is the product developer's customer in either the food service or food ingredient industries? In the food service industry, the customer is a manager or franchisee in a fast-food chain, a chef–owner of a restaurant, or a manager of a commissary of a hospital. In the food ingredient industry, a product developer of ingredients sells to another ingredient manufacturer (the customer) or to a company manufacturing some added-value product. These customers are not the ultimate consumers, but like consumers they have needs.

This two-tiered user category introduces novel problems. How does one screen a product, such as a new ingredient, when one does not know specifically how or in what product that ingredient will be used? How does one get market research, conduct a test market, or evaluate a test market? Developers of products for either of these two markets must satisfy the needs and expectations of customers directly and the needs of consumers indirectly. The food ingredient industry will be dealt with in the next chapter.

I. CHARACTERISTICS OF THE FOOD SERVICE MARKET

In the past, most meals were eaten at home and generally by people in a family unit. More and more, people are changing their eating habits; the traditional family unit has changed with the changing lifestyles of consumers. Today, fewer meals are eaten at home and rarely in a family unit, except for festive occasions. The result is an increase in meals-away-from-home.

Food service is a large and growing market sector of the food industry. A major characteristic is its highly fragmented nature, as the following examples suggest:

vending machines	gasoline stations
hospital catering	grocery stores
supermarkets	fast-food restaurants
family restaurants	white-tablecloth restaurants
transportation	industrial work sites
hotels/motels	military
penal institutions	school meal programs
catering services	bars, pubs, nightclubs
street vendors	

The categories on this list, which is by no means complete (see Bolaffi and Lulay, 1989, for a more extensive listing), could each be further subdivided. For example, fast food restaurants could be separated into independent family-owned businesses, large multiunit chains, restaurants with sit-down service, or those with only counter or take-out service. Likewise, industrial work-site feeding could be an urban or suburban factory-run cafeteria offering a hot lunch, or a cafeteria serving a 24-hour-a-day work force, or a food service organized for an isolated oil rig in the North Sea. Hospital feeding can involve nonspecial diets in convalescent homes or special diets for those with impaired digestive systems or other disorders. Each category of food service has its own peculiar needs.

Food service can be roughly partitioned into two sectors: the commercial sector and the noncommercial sector. (The term "noncommercial" is not the best descriptive term.) The commercial sector runs the entire gamut of the restaurant industry, from the stand-alone white-tablecloth restaurant to the vending machine dispenser found in hotels, motels, schools, workplaces, or shopping malls. Also included are deli counters in food stores (a challenge to the fast-food restaurants), meal trucks servicing factories and construction sites, street vendors with their various traditional finger foods (fish and chips, hamburgers, hot dogs, tacos, and gyros) and not-so-traditional ethnic finger foods.

Another name given to this sector is the "free choice" market. That is, consumers have the option of going to a variety of outlets if choices at one do not appeal; they are noncaptive. Hence, this market is also referred to as the noncaptive market. The meaning becomes apparent when the noncommercial market is described.

In the noncommercial market (also called the institutional market), consumers have no freedom of choice in the menu or only a very limited selection (i.e., the familiar "Will that be chicken or fish, Maam?" heard on airlines). Consumers are captive: they cannot go elsewhere to eat. Patients in hospitals and convalescent homes may require very restricted diets that cannot cater to their personal choices. Situations such as the following all represent markets in which captive consumers are presented with limited or no-choice menus:

- school meal programs,
- company cafeterias in isolated industrial parks,
- meals during air, bus, or train travel,
- meals in prison and other rehabilitation centers,
- military feeding while on maneuvers in isolated situations (arctic survival training, air and naval patrols),
- relief feeding in famines, natural disasters, etc.

Some authors (Bolaffi and Lulay, 1989, for example) separate out the military food service as distinct, by virtue of the product standards, labeling, and packaging requirements, as well as the bidding and tendering requirements of the government. However, for the purposes of this discussion, military feeding will be included with the noncommercial or institutional sector.

Both captive and noncaptive sectors of the food service industry have unique problems, which make generalization about them difficult. They both suffer shortages of quality labor, which are exacerbated by the rising costs of labor. Labor costs vary widely. Pine and Ball (1987) found that wage bills (as percentage of sales minus pretax net profits) ranged between 8 and 38%, depending on the class of the establishment. Those that are labor-cheap were largely in the catering-only end of food service, i.e., putting together preprepared items, or they were businesses lacking a personal service aspect (transportation catering). These figures apply to the U.K. industry but are generally applicable in other countries. Thus, even within the food service business, there are wide variations in labor costs. The fast-food sector has had to encourage nontraditional sources of labor, with an appeal to seniors such as retired pensioners.

II. CHARACTERISTICS OF THE COMMERCIAL SECTOR

Diversity is the key descriptive characteristic of the commercial sector. Product development here is challenged by the diversity of facilities and of equipment in these facilities. This diversity also runs the gamut of labor skills, from chefs in top-of-the-line gourmet restaurants to street vendors in their peddle carts selling hot dogs, hot sausages, and sauerkraut.

There is diversity in the different quality expectations of consumers, which will confound product development for this sector. Consumers in high class restaurants are looking for service, atmosphere and relaxation, quality of taste, and presentation. They are less likely to be concerned with price *but will demand quality and service for the price.* Consumers buying from street vendors are looking for cheap, good, wholesome (safe) food to quickly satisfy a hunger while on a lunch break or between breaks in outdoor play.

A major characteristic of the commercial sector is that it is profit motivated. Plainly and simply, this sector must be profitable. It is most unusual to see a commercial eating establishment run at a loss, unless some person or organization is subsidizing the operation. Only two examples of this type of operation come to mind: a demonstration restaurant run by a government- or university-supported hotel/restaurant training school and a restaurant at a private golf club subsidized by members' fees.

A. Food Preparation and Storage Facilities

The food preparation and storage facilities of restaurants must rank as the most diversely equipped facilities known. White-tablecloth restaurants, family-style restaurants, ethnic restaurants, hotel kitchens catering to two or three outlets within the hotel complex, fast-food franchises, mom-and-pop diners, or bars serving hot hors d'oeuvres will have food preparation equipment ranging from warming units to steamers to fryers to specialty wood stoves to tandoors to woks to hot stones to 50-gal steam-jacketed kettles to mixers and microwave ovens to potato peelers, grinders and slicers to walk-in freezers and refrigerators. Some will be well equipped; some will have less equipment than a well-run family kitchen would have.

Many if not most restaurant kitchens are run at one temperature of heating for all menu items served. This puts serious constraints on designing products for development in this sector. Products developed for restaurants operating this way must withstand this temperature stress and maintain their high quality during preparation. Product developers must design (formulate) product or ingredients that have these attributes.

Storage is limited in food service establishments. In some outlets, such as bars, there are basically no storage facilities. These establishments rely on same-day deliveries. The caterer arrives with hot finger food and warmers, sets up in the bar for happy hour, and returns to remove the dirty ware. For others, storage facilities can range from the adequate but limited for full service restaurants, to the very limited, barely adequate for street vendors.

Limited and inadequate storage facilities present challenges to product developers in another way. Stabilizing systems designed into the products during development must anticipate the challenges in the distribution, the storage, and the vending environment that these products might encounter. This is particularly so for products for food service companies supplying convenience stores, vending machines, and street vendors.

For these latter, distribution (stock rotation) and stability are very important factors in their survival and a major problem for their suppliers. For example, in the vending machine sector, there have been several developments: newer machines incorporating microwave ovens or deep-fat fryers make a wider selection of product offerings now possible; combination plates, sandwich, salad, and fries or soup, sandwich, and dessert, are now regular offerings (Williams, 1991). These new menu items are much more sensitive to time and temperature abuse.

If storage for food supplies is limited, then storage for waste is even more at a premium. Products for this industry must be designed to produce as little waste as a byproduct as possible. Prepreparation is a key added-value feature. For take-away hot foods, fast-food outlets face another problem: packaging. It must be environmentally friendly.

B. Labor

In fast-food restaurants in particular, but also in many other food service establishments as well, labor is largely unskilled, frequently unreliable, and demonstrates a high rate of absenteeism. Personnel are usually in transitional stages in their careers. That is, they are teenagers or college students supporting themselves through school, the temporarily unemployed, or those supplementing family incomes. They are usually not experienced in food service methods, are not careerists in food service, and want to move on at the earliest opportunity into their chosen careers. Training such candidates can be discouraging and fruitless for food service employers.

Therefore, preparation of products to be used in food service must be kept simple, must require the minimal amount of labor, and must produce the least amount of waste for handling. Products for food service use must be "error-proof". Product developers, in addition to the task of making certain that a product is safe, have the added tasks of making certain that its storage presents no complications to its safety or quality and that preparation in the hectic environment of a kitchen is foolproof.

On the other hand, quality restaurants have skilled personnel, chefs as well as sous-chefs, available. These can take prepared mixes (developed products to be used as ingredients) and use their skills to prepare and present attractive, added-value products. Likewise, these chefs can take unused portions and reuse these, safely, in other quality dishes. Yesterday's unused poached salmon can be today's salmon mousse. Such skills reduce wastes and thereby minimize cost overruns. This is not the case in fast-food restaurants, where such skills are generally lacking. Nothing of chance respecting safety can be left to personnel. Unused food or warmed and thawed portions must be garbaged.

C. Consistency: Price, Quality, and Safety

Price, consistency of quality, and safety for high volume products are particularly important factors to be considered in fast-food restaurants, especially the franchised chains. Price is regulated by what the trade will bear and what the target consumer for whom the outlets are aiming will pay. Price also enters into portion control. One cannot have patrons observing others and thinking, "Their portions are bigger than mine!" Portion control, therefore, becomes a very important element of price and service.

Quality is equally important. Quality must be consistent with the price the consumer is willing to pay for that quality at each and every sitting. In addition to this, sensory qualities of both food and service must not vary from restaurant to restaurant in the same chain.

Safety, that is, considerations for food poisoning and intoxication, is a major concern in the food service industry (Snyder, 1981; Snyder, Jr., 1986). Food poisoning outbreaks, with their attendant media coverage, can destroy a restaurant. Solberg and his colleagues (1990), in an extensive survey of meal items and food preparation facilities at Rutgers University's New Brunswick, New Jersey campus, however, found food pathogens to be a minor occurrence in menu items. Protein salad foods presented the greatest safety concerns.

On the other hand, food allergies and the suffering and even deaths of consumers as a result of allergic reactions to foods eaten in food service establishments have also emerged as a problem (Williams, 1992). In one case, a teenager ate an apple turnover purchased from a fast-food restaurant and died. The youngster knew he was allergic to nuts but was unaware that the filling of the turnover contained ground hazelnuts. Such deaths may seem rare and unusual, but allergy specialists in Canada estimate there is at least one death a month due to allergic reactions. In the U.S., with a tenfold higher population, this might suggest ten or more deaths a month due to eating prepared foods.

There is sufficient concern on the part of food service establishments to post allergy charts, listing the ingredients in the menu items served in the restaurant (Anon., *Can. Consumer*, 1988). The consequence of this precaution forces manufacturers supplying food service establishments to be certain that the composition of all ingredients they use in their products fit these charts. This requires vigilance on the part of manufacturers as they switch ingredient suppliers looking for better cost control.

Safety of prepared foods and ingredients poses a severe challenge to the food service industry and to suppliers to this sector.

III. CHARACTERISTICS OF THE NONCOMMERCIAL SECTOR

In the preceding section, it could be seen that consumers dominate the commercial (noncaptive) food service market. Consumers can choose a variety of dishes from several price ranges on the menu or, not wishing to do that, have the choice to go elsewhere. The restauranteur/owner or manager of the commercial food service outlet observes and caters to consumers' needs. The day's cash receipts, the chef's inventory of supplies, or the frequency of particular meal selections recorded from the receipts provide evidence of what consumers are choosing.

In the noncommercial sector, captive consumers have little or no influence in the food choices available to them. The operator of the establishment determines both the level of service and quality of the food within constraints of budgets established by ticket sales, government subsidies, private fees, or charitable donations. Consumers are confined; they cannot walk to another eating establishment.

However, it must not be thought that the consumer does not have any influence. Drastic action on the part of the consumer can and does bring about changes in both service and quality. But the action is always drastic and dramatic. Poor quality prison meals have been known to spark riots and demonstrations. Wastage in the garbage cans in school cafeterias and tales from children to their parents can be very telling commentaries on the quality of food in school lunch programs. Those who offer poor airline and train food can soon find travelers switching travel plans to more accommodating carriers. Where meals are part of the fringe benefits of a union contract, the choice and quality of food available can certainly be factors in union/management strife. One chuckles today at notices in coastal logging and fishing camps in bygone days that promise that salmon and lobster will not be served more than three times a week.

Noncommercial outlets suffer from many of the same problems that plague the commercial sector of the food service industry. In addition, there are unique problems not seen in other sectors. For example, military feeding presents unusual conditions in peace time and in war time. Peace time maneuvers in tropical or arctic terrain or hostile enemy activity in war time make feeding situations difficult. Also, hospital feeding presents a very different feeding situation. It includes both the commercial public cafeteria for hospital staff and visitors and the institutional sector, where very special meal requirements are necessary for patients.

Unlike the commercial sector, the noncommercial sector food service is not profit-driven. This should not be understood to imply that this sector is run at a loss or that there are not attempts to make it profitable. This sector operates within a budget, and staying within budgetary guidelines is what drives it.

A. Food Preparation and Storage Facilities

In general, food preparation and storage facilities in noncommercial establishments are better than or more adequate to their tasks than those of commercial establishments. They are usually run by professionals trained in food service or are catered to by industrial caterers. While they are on limited budgets, they frequently can purchase in bulk and obtain reduced prices.

Some U.S. hospitals have rented out part of their food service facilities to fast-food chains. This presents an interesting combination of commercial and institutional (noncommercial) food service in one facility.

B. Labor

The same problems that befall the commercial sector respecting labor befall the institutional sector. Labor is transient and largely unskilled, with a large incidence of absenteeism. Workers are often newly arrived immigrants. There can, therefore, be communication problems if there are language barriers. In some penal and correctional institutions, inmates may serve as help.

There are, however, wide variations in skills in this sector. Dieticians required in hospitals and nursing care facilities and military chefs are highly trained; school lunch programs are usually under the supervision of equally skilled personnel. Nevertheless, wages are at the minimum level for unskilled labor, and at the bottom end of the wage scale, job security is absent. These are not elements conducive to attracting skilled people. On-the-job training is difficult in this environment. The exception is in the military, where training programs have proven very effective in raising the skill levels of food handlers and cooks.

C. Consistency: Price, Quality, and Safety

Price and quality are dictated by budget constraints, but each food service facility within the noncommercial sector has very different standards for nutritional quality and safety.

The nutritional content of foods served, for example, is not a critical issue for airlines, where consumers require only a snack, a light meal, or a hot meal during the short duration of the trip. The long-term health and welfare of airline passengers are not dependent on the food served.

On the other hand, hospitals and other nursing care facilities, agencies providing home-delivered meals to the elderly, government-supported programs such as school lunches, and the military have a very special concern for the nutritional content of their meals. Consumers of these meals are long-term, and the nutritional adequacy of their meals has a direct bearing on their health and safety.

Hospitals and hospital care facilities must also cater to special diets required by patients recovering from the trauma of an operation or from treatment of various medical problems. Therefore, food must be attractive and flavorful to encourage eating by these patients, and, at the same time, food must be prepared to satisfy the needs of staff, students, and visitors.

All establishments are concerned with the safety of food served. However, in facilities such as health care establishments and military feeding, safety of the food becomes extremely important. The weakened condition of patients and the fact that many of the patients may have deficient immune systems emphasize this extra need for safety. Microbiological loads that would normally be tolerated with no difficulty by healthy individuals may fell these patients. An army cannot come down with some foodborne illness.

Confounding all the above is the lack of standards for special dietary foods. That is, there are no standards for nutritional foods, for soft foods, for low-salt foods, or for any of the special conditions dictated by convalescent diets that are required. Consequently, developers wishing to put products into the special market niches have no guidelines. They must, therefore, work closely with dieticians in development.

D. The Military Sector of the Institutional Market

It is important to any country to keep its soldiers healthy and fit. The dictum, "an army marches on its stomach", attributed to Napoleon Bonaparte, was never truer. Food, good food, keeps an army going and its morale high. This requires good quality, flavorful, capably prepared food.

The military market has unique problems and opportunities to present to new food product developers and food preparation equipment suppliers. This market is also highly fragmented. Food service facilities and operations range from officers' messes to NCO messes to combat field kitchens to vending machines and snack bars to hospital care facilities. Most of these facilities must be operated 24 hours a day to accommodate the shift work involved.

In addition, there must be special attention paid to the nutrient content and nutrient density of the food supplied because of the very active lifestyle. Military personnel must perform in high tension, strenuous conditions sometimes for long periods of time. In covert action or in combat zones, food must be carried conveniently by the individual combatant while still with full gear. It must be prepared with minimal equipment yet provide ample nutrition.

When and where this food might have to be prepared and consumed introduce problems. Ease of preparation is important because food may be prepared or reconstituted in less than ideal conditions in combat zones, in cramped conditions in submarines, or on airplanes and often by the unskilled personnel themselves. Packaging must be light in weight and yet provide protection for the food from all environmental conditions (which are unspecified) and from any treatment (including air drops) it might endure. Yet the package must be easily opened and readily disposed of, lest obvious waste disposal dumps be seen by enemy aerial observers as a sign of a kitchen nearby.

As is the case with all government procurement purchases, products must adhere to rigid standards and specifications respecting ingredients, processing, and packaging. These standards and specifications can be quite detailed. Suppliers are advised to obtain copies of them before attempting to manufacture products.

A possible cause for failure of new products, discussed earlier, could be the introduction of products into markets dominated by a single customer; the military represents such a market. On the other hand, the military market (and the food service market, in general) does present unique opportunities to introduce new products and develop within the population a liking for, a familiarity with, in other words, an education for novel food products.

E. The Health Care Sector of the Institutional Market

This sector includes hospital feeding and food service facilities, convalescent care feeding, and feeding of the elderly, whether they be in their own homes, in residences for the mobile elderly, or institutionalized in nursing homes. This partial listing of health care facilities should indicate that this sector is itself highly fragmented.

Meals with a quality and service equivalent to any limited menu whitetablecloth restaurant can be home-delivered to the elderly and to residences for the elderly (where the elderly are, in both instances, quite mobile). Agencies can present products of superb and high nutritional value.

As in military food service, there is a gamut of eating, dining, and snacking opportunities. Unlike the military food service, procurement and purchasing procedures are less standardized, and product specifications are more flexible (not lower).

Regional districts frequently control or influence the buying power of hospitals. As a result, their purchasing power can vary widely. Meal items are often purchased from privately run food commissaries. Nevertheless, many hospitals do still prepare their own meals or prefer to prepare their own special dietary meals because of the lack of accepted industrial standards (Burch and Sawyer, 1986).

Hospital food service presents many unique problems and opportunities for product development (Matthews, 1982; Burch and Sawyer, 1986). For example, there is a public cafeteria that provides food services for the doctors, nurses, students, and staff, as well as for the general public who are visiting patients. This operates 24 hours a day to feed all shifts of supervisory and maintenance staff. But the unique aspect of the operation in any health care facility is the challenge to meet the nutritional and dietary needs of patients who, because of their ailments, may have a wide variety of dietary requirements superimposed upon numerous dietary, religious, or ethnic taboos. Health care facilities cater to bland diets, low-fat diets, calorie-controlled diets, low-fiber diets, soft diets, sodium-restricted diets, and diets for diabetics.

Food preparation systems in use in U.S. hospitals employ one of five common preparation systems (Matthews, 1982):

- a cook/serve system, whereby food is prepared on site, plated, and distributed on trays to patients;
- a cook/chill system, in which the food can be prepared either on- or off-site ahead of use. After preparation, the food is rapidly cooled, held refrigerated for not more than a day (but see Mason et al., 1990), plated, and distributed to patients from specially designed heating carts;
- a cook/freeze system, which is similar to the preceding with the additional step requiring the thawing of menu items;
- a thaw/heat/serve system, where the main menu item is precooked and frozen either on- or off-site. On-site, the product is thawed, heated, and plated;
- a heat/serve system, using the thermally processed steam table tray containers. The main menu item can be heated and held hot before plating.

It should be noted that these systems are not unique to the health care food service industry and are used in military feeding (especially the heat/serve system) and in many commissaries servicing a variety of cafeterias and other dining establishments. Awareness of the rigors that such systems impose on food products and ingredients is necessary before developers can design products that withstand these rigors.

Mason and co-workers (1990) and Livingston (1990) discuss commercial food preparation systems used in hospitals and other food service outlets and evaluate these with respect to their influence on quality of the food.

IV. CUSTOMERS AND CONSUMERS: WHAT ARE THEIR NEEDS?

A. Customers and Consumers

The distinction between customer and consumer, described and adopted earlier in this chapter, must be kept clearly in mind in any discussion of new product development for the food service market. The manager/owner/chef (the customer) of any food service outlet buys products to fashion into menu items for consumers. This customer has needs and desires (quite apart from the very natural desire not to lose consumers through poor food!) that must be satisfied by the prospective developer of new products. If these needs and desires are not satisfied, a customer is lost, and ultimately many consumers are lost.

The needs of two distinct user entities must be catered to for successful product development. They represent two separate market niches with different requirements within food service.

B. Physical Facilities of the Customer

New products for the food service industry must be capable of being prepared with the equipment that customers have on their premises. This equipment can be very primitive or very sophisticated, and the food product developer must recognize the restrictions in product design that the state of the kitchen preparation equipment may pose. Not only that restriction applies, but the product must be platable with the smallware that is available.

Settlemyer (1986) describes the development of a buttermilk biscuit for a chain of restaurants. During development, it was determined that although ovens suitable to bake the biscuits were available in individual outlets, warming ovens to hold the biscuits were not. Thus, included in the development process was the need to research different brands of biscuit warmers.

One fast-food, multioutlet company solved the problem of compatibility of products and equipment. This company had a fully operational outlet at its development center, not merely a mock-up test kitchen. The sole purpose of this outlet was to provide a one-store test to determine how operationally compatible products were before any test market was attempted (Peters, 1980). In both Settlemyer's and Peters' examples, the needs of customers (chef/ owner/managers) were catered to first.

Since dry goods storage space and refrigerator and freezer space are limited, new products must be evaluated for the effect they will have on available storage space. Unused product must be easily resealable and capable of being safely stored.

C. Energy Requirements

Types of energy used in food service facilities and their costs vary widely within any country and from country to country. Energy as a cost factor in food service operations is, as a consequence, highly variable. Where prepared foods received from a central commissary are used, less energy may be required in preparation and presentation. *De novo* preparation from raw ingredients would require more energy for food preparation and cooking. Energy is, along with labor, a considerable contributor to overhead expenses. Efforts to reduce the energy used in meal preparation would be greatly appreciated by food service operators. In military feeding, in particular, energy conservation would be very welcome in food preparation, since energy sources must be moved with the marching, sailing, or flying consumers.

To design products that conserve energy in their preparation, developers need to determine how and where energy is required in food preparation. If only the prepared food itself is considered, energy is absorbed by the average temperature increases in foods as the masses of foods are brought to the necessary preparation and serving temperatures and phase changes in foods of mixed composition as solids are converted to liquids (dissolving and melting) or water to steam (concentrating) (Norwig and Thompson, 1984). Norwig and Thompson describe the calculation of energy requirements using the frying of french fry-cut potatoes and the cooking of frozen hamburger patties as examples.

Snyder (1984) suggests that foods, particularly sauces and gravies, can be formulated with the correct amount of liquid incorporated initially to get the desired finished consistency. This would reduce the later need to boil the sauces down to evaporate water. No concentration step would be required, and both time and energy could be saved. An additional benefit would be the improvement in quality, since less heat damage to sensitive ingredients in the sauce would occur.

Many factors affect the energy absorption (and quality) of food products during heating: shape, thickness (McProud and Lund, 1983; Ohlsson, 1986), density (McProud and Lund, 1983), composition (Bengtsson, 1986; McProud and Lund, 1983), and method of heating by microwaves (Decareau, 1986), by convection or radiant ovens (Skjöldebrand, 1986); by boiling (Ohlsson, 1986); by frying (Skjöldebrand, 1986), or by grilling (Bengtsson, 1986).

It should be pointed out that energy usage is not confined to food preparation in food service establishments. Energy is consumed in lighting, heating, air-conditioning, and fans in both the dining and in the kitchen areas. In addition, refrigerators, freezers, serving cabinets, and warming ovens, which have nothing to do with preparation, consume energy.

Designing products or ingredients to be energy efficient is an important consideration in products for the food service industry. Food service operators have a need to reduce their overheads; reducing energy consumption is one way to do this.

D. Labor

Labor, its availability, its skill level, and its cost, are problems for managers of any food service outlet. Preparation must, therefore, be simple.

Developers need to provide clear and explicit instructions for:

- storage of the product and unused portions of the product,
- preparation of the product itself and preparation of any of its recipe variations (multiple uses of products or ingredients),
- the safe display of the product,
- serving the product.

The product must economize on labor in its preparation and must be capable of preparation in the rushed, hot, steamy, and temperamental atmosphere of the kitchen. Preparation may not always be by chefs but frequently by young, unskilled and untrained, part-time staff working at a job, not at a career as a cook.

In fast-food restaurants, preparation must not only be simple and uncomplicated to produce products with uniform quality, but preparation time must be short. Where the food service outlet caters to a high-volume turnover of consumers, rapidity of preparation is important.

Simplicity of preparation, versatility of usage, as well as rapidity of preparation are needed features for food service operators. These are worthy considerations for design into new products and ingredients.

E. Waste Handling

Clean-up and waste disposal associated with the use or preparation of any new product must be simple. With storage already limited for supplies and food, storage for waste must be held to a minimum. The use of new products or ingredients must result in as little waste as possible being produced. Unused portions must be easily and safely stored without special storage requirements.

There is another consideration — sanitation. Any product introduced into the food service establishment must not introduce any unusual hazards nor require unusual or extraordinary handling techniques with respect to hygiene, clean-up, and sanitation.

F. Consumers: General

So far, only the needs of the customer have been considered. It can be seen that some of these needs and wants are very specific. A very basic point about product development in the food service market must be made. It is this: it is no longer enough, if, indeed, it ever was, to make a product and put that product into an institutional-sized container. That is not product development in this unique market. Products designed for home use do not become products suited to the needs of food service managers merely by using larger-sized containers.

To enhance the chances of being successful, product developers should speak to food service operators, discuss with them who their target market is and what their requirements for new products are, and be guided by these in new product development. Working closely with the customer permits the developer of either new products or ingredients to better focus research to design products more suited to the physical and operational needs of customers. Unless food developers know what the ultimate preparation in the kitchen and presentation to the consumer will be, developers will continue to have difficulties designing suitable new products for this market.

With the food service industry, catering is unique, in that in addition to providing an *eating occasion*, with the emphasis on occasion, there is a sociability aspect attached to that provision. "Catering systems are socio-technical systems. Catering systems that are centered wholly on the technical aspects and ignore the social aspects will fail ... change in production methodology or product formulation and presentation must recognize the social context within which the final product outcome is to be consumed" (Glew, 1986).

Whether consumers "dine out" or "grab a bite", developers must recognize that these meals are being consumed in a social context. Food must please; food must entertain; food must satisfy; food must comfort.

What does this conjure up for new product developers? Fun foods? Comfort foods? Entertainment foods? The context of food usage, the occasions during which it is consumed, must be understood as important factors in both development and marketing. There is an entertainment/sociability/warmth factor that must be designed into these foods. Food retailers attempt to capitalize on this aspect of food with in-store bakeries wafting the aroma of freshly baked bread throughout their stores. The smell of freshly baked bread has a warmness to it; it evokes memories. It is a smell that was most missed by soldiers in combat situations.

G. The Consumer: Nutrition

Today's consumer is concerned about nutrition. But how deeply this concern extends when the context of eating is social is very difficult to assess. Nevertheless, customers who recognize the consumers' desire for cutting down excess fat, for getting plenty of good fiber in foods, and for consuming minimally processed foods are driven to devise products that satisfy these needs. They devised reduced-fat burgers and veggie burgers. Yet quietly, reduced-fat burgers have disappeared from the marketplace. Is the entertainment/sociability/warmth factor associated with eating occasions in consumers' minds greater than their nutritional concerns?

Many of the fast-food chains have extended their menus to present healthful variants of standard products or to present new eating opportunities; e.g., breakfast menus have been developed and soup and a variety of salads are being offered. Extended menus require new ingredients and new food products. Where new products can serve a multitude of uses in a food service outlet, such as to satisfy the consumer with a variety of tasty products and the customer by reducing the variety of ingredients carried, everybody wins.

Nutrition, price, taste appeal, and convenience — elements satisfying to both customers and consumers — influence acceptance of new products in food service markets. Nutrition is a factor in acceptance but not the only one, nor, in my opinion, the most important one. Consumers have been bombarded with low-fat, low-calorie, low-sodium, sugar-free, high-fiber, vitaminized foods on both store shelves and in food service outlets, and they are confused. The trend to healthy eating will remain, but this market must be allowed to develop at its own pace.

V. QUALITY IN THE FOOD SERVICE MARKET

There are two judges of quality: customers and consumers. Quality is first determined by the management of the food service outlet, be it a commercial or a noncommercial outlet. If quality is to be understood as the satisfaction of needs, then product developers must first meet the needs of the kitchen staff.

Here, quality attributes must include price, convenience, minimal labor requirements, short preparation time, consistent quality, individual packaging or controlled portions, and, if possible, multiple uses for the product. When these are satisfied, then perhaps the product will be put to the consumer for judging.

The ultimate judge of quality is the consumer. Criteria for assessing this judgment are the trash bin and loss of sales.

A. Safety

Concerns for safety with respect to both hazards of public health significance and hazards of economic significance are paramount in food service. Programs, e.g., HACCP programs, used to resolve these concerns do not differ greatly from those discussed for product development earlier. No hazard of public health significance must be associated with new products or ingredients.

Extra consideration for the safety and stability of foods and ingredients must be given because of the special stresses that are normal in the food service industry. The limited storage area, the frenetic activity in the preparation area, and the need to display or hold product hot in serving areas present additional challenges to the quality characteristics of products or ingredients.

A further consideration regarding safety stems, as mentioned before, from the fact that in health care feeding, many patients have compromised immune systems, and special care must be paid to the microbiological safety of ingredients and products intended for their use.

In the commercial sector, allergic reactions of consumers to products or their ingredients have prompted sufficient concern that many food service outlets insist that products be, so far as is possible, free of allergenic ingredients. They post lists of ingredients used in their products. This puts an onus on suppliers and would-be suppliers, requiring that suppliers to this industry adhere strictly to approved formulations, substituting no alternative ingredients. In their turn, they must insist that the ingredients they receive have fixed compositions, with no substitutes, and they must insist on a strict truth-in-labeling policy with all their suppliers.

B. Nutrition

Nutrition as a quality feature is two-faceted. There are both customer needs and consumer needs, particularly respecting health care feeding. Customers in the health care field (dieticians) require unique products to satisfy special requirements of postoperative patients, patients undergoing cancer therapy, or those under drug therapy. They also require very specific nutritional information for consumers (patients or the elderly enjoying a meal service) in their care. They need much more detailed, indeed, esoteric nutritional data than that required in most nutritional labeling regulations. Developers must be prepared to provide this information in order to assist dieticians in the preparation of the many different dietary regimens required for patients. Schmidl and colleagues (1988) provide an extensive review on parenteral and enteral feeding systems, a rather exotic branch of hospital care feeding.

Is there a problem of nutritional quality with the food service most used by the consuming public, the fast-food sector? Fast-foods have not received good publicity concerning the nutrition they offer consumers. For example, Hibler (1988) reported the results of a sampling of burgers, fries, and chicken, highly ranked favorites among frequenters of fast food outlets. The calories per gram of burgers ranged from 1.9 to 2.9 cal/gm; of chicken entrees (nuggets or fingers) from 2.4 to 3.1 cal/gm; and of fries from 2.7 to 3.5 cal/gm. If the weight of entrees were constant (they were not), hamburgers should be the least calorific choice. The calories from fat were highest in the chicken entrees, ranging from 39 to 55% (average 49.5%) and were almost identical in the burgers (36 to 50%: average 44%) and the fries (37 to 48%: average 42%). A typical burger and fries or chicken pieces and fries would put the percentage of calories from fat well over the recommended level of 30%. Ryley (1983) calculated that, even back in 1982, fast foods and snacks contributed over 16% of the daily fat intake per person in the U.K.

There is a concern with nutritional quality in the feeding of the elderly at home. Turner and Glew (1982) studied the nutrient content (protein, energy, calcium, iron, and ascorbic acid) of meals delivered in Leeds (U.K.) and provided by six food service organizations. They found significant weight differences between the meals supplied by the different organizations, as well as between the protein contents of the meals. Meals supplied between 20 to 48% of the recommended energy intake for elderly people, which represents under to grossly over the energy requirement for the main daily meal. Of particular importance to geriatric nutrition, it was found that the calcium content of the meals varied widely from wholly inadequate to ample, dependent primarily on whether the dessert was milk-based or not. Iron content was found to be just adequate for the elderly, but ascorbic acid contents varied widely, from providing more than 50% to less than 25% of the recommended daily intake. However, with ascorbic acid, significant losses were noted between the first and last meal deliveries, as might have been guessed from its lability. Retaining heat during transportation, as well as the damage that the duration of hot transportation itself had on vitamin C, were weak links in delivering nutrition to the elderly.

Some detail has been expended in reviewing these studies on the nutrition of food service meals to point out a major quality problem for developers of products, whether they be *sous-vide*, frozen or other chilled food entree items, or dry powder bases for soups, gravies, or desserts. What are the nutritional standards or guidelines for these products? There are no nutrient standards or even nutrient specifications for products meant for institutional or military or school meal programs. And there certainly are no standards for the special diets (What is soft? How soft? and How does one measure "soft", by fiber content?)

required by health care establishments. There are excellent opportunities in which industry could assist these establishments and find good and profitable market niches for their products. Developers could create added-value menu items with established nutrient content (perhaps one-third the daily requirement) for health care establishments, or semiprepared foods with fixed soluble/insoluble fiber content ratios.

VI. DEVELOPMENT OF PRODUCTS FOR THE FOOD SERVICE MARKET

Development for the food service market follows pathways similar to development in the food retail sector. This holds true whether a company is attempting to introduce new products to operators of food service establishments or whether a food service company itself wishes to bring new items onto its menu (Peters, 1980; Settlemyer, 1986). Both should have first developed objectives that they wish to attain. From these, there should arise strategies and then tactics to reach these objectives within the time and financial constraints that management imposes.

To become successful suppliers to the food service industry, companies must work closely with operators or at least be very knowledgeable about the catering industry. This market is highly fragmented. Each fragment represents a marketing niche in itself.

Development teams require the following:

- a very clear and specific statement of what the product is, i.e., dessert, main course, sidedish, etc.;
- identification of which targeted consumer the product is meant for;
- knowledge of which meal the product is intended for;
- the price range of the product.

Price becomes a somewhat more important factor than it was in the retail food market. Price must fit into the price structure of a whole meal. Price is much more obviously comparable to the other items, soup to salad to entree etc., on the menu. In a grocery purchase at a supermarket, any one item can be lost among all the others; its price does not stand out so noticeably as a proportion of the total purchase. Not so on a menu. Here, price sticks out like the proverbial sore thumb.

There follows the usual routine of consumer research with focus groups, questionnaires, and interviews to get a clear and comprehensive reaction by the targeted consumer to the product concept. If, for example, the targeted consumer is the "wellderly", that is, the over-50 market, then that segment of the market requires careful study. Physiological changes in these older consumers open opportunities for healthful new products designed especially for them. Designing meal occasions and menu items for them to meet their needs can be challenging. Food service is a highly fragmented market to which developers must adapt with new products and menu modifications.

If competitive products are on the market, an audit of these products will be imperative to provide some idea of the quality levels in the marketplace, the pricing structure, and consumer reaction to them.

Criteria in screening are:

- price and profitability (including labor-saving in preparation, rapidity of preparation, serving time, as well as portion control);
- constraints imposed by preparation and display equipment limitations;
- market reactions, i.e., consumer acceptance of the product and consumer reaction to what was displaced on the menu; and
- the safety and quality of the product with respect to its stability in view of the stresses of that sector of the food service market for which the product is intended.

Since many products are prepared in a central commissary that may be anywhere from several floors away in a hotel to several hundreds of miles away, in the case of fast-food chains and health care feeding establishments, the potential for abusive mishandling in distribution is a very important screening consideration.

Developers of food service items need to provide a very clear list of instructions for the storage of the product, its preparation, whether it is capable of multiple use, preparation of all its variants, its display, the method of serving, and storage or treatment of unused portions. This is a requirement less essential in retail food product development.

At this point in time, a consumer test can begin. This can be a small test of the product in two or three units, or it may involve a dozen or more food service outlets. Marketing support can vary from simple table tents to television, radio, and newspaper advertisements supported by coupons, in-store displays, and free sampling.

VII. CRITERIA FOR EVALUATING A TEST MARKET

Consumer reaction to new products can be found by use of questionnaires (consumer intercepts). Cash register data can provide information on the new product as a percentage of sales. More dramatically, analysis of the amount of new product found in the trash containers at test sites provides ample evidence of acceptance. A careful interpretation of all data will be necessary to provide information that may herald a successful introduction or prevent a commercial disaster.

In the first instance, questionnaires permit the developer to evaluate the consumers' reactions to the new product and, if evaluation warrants, set about refining the product to better adapt to consumers' needs.

In the second instance, additional information can be had from cash register data. The impact of the new product on improving sales or the impact of its introduction on the cannibalization of some other menu item can be determined.

Finally, there is the physical evidence of rejection by the consumer: waste. There are three potential sources of waste in any food service establishment. There is preparation or kitchen waste, which can be classified broadly as food purchased for kitchen use but discarded during preparation or spoiled in storage. Kitchen waste will, of course, be higher in establishments doing their own preparation work and not relying on preprepared foods. The second source of waste, service waste, is prepared food left in warmers or steam tables and not purchased or accepted by consumers. The amount of kitchen waste and service waste is largely although not entirely a measure of the management skills of the establishment. Finally, there is consumer waste: food purchased by the consumer but discarded. This is a measure of rejection.

How waste is to be measured and assessed presents some problems. Banks and Collison (1981) studied the problem of waste in 39 catering establishments in the U.K. and discussed the factors affecting waste, not the least of which is the size of the meal. Lack of attention by the establishment to portion control increased waste, but the amount of convenience food used by the establishment decreased it.

Kirk and Osner (1981) agree that consumer waste can be a sign of poor portion control. They state:

> It may be thought that plate waste does not represent a financial loss to the establishment since the food has been paid for by the consumer. However, poor portion control can lead to more food being produced than is required or to a loss of potential sales.

In addition to poor portion control, there is another factor contributing to plate waste. People have varied attitudes towards the edibility of particular food items, for example, potato skins, the skins on other vegetables such as cucumbers or zucchini, and giblets.

Nevertheless, analysis of consumer waste can be a useful tool in assessing consumer acceptance of menu items. Its interpretation must be used cautiously. Cash register receipts provide an indication of purchase, but the garbage bin audit can tell of the acceptance of the new product.

Consumer research requires careful assessment. Does the data represent only the regional preferences of the test area selected for introduction, or can the data can be extrapolated to wider market areas? Regional dishes in one area of the country may not be equally well accepted in other areas. Impartial answers to these and other questions are required.

Such consumer feedback can then be directed to determining what necessary modifications are required in the product to satisfy the needs arising from the operation of the outlet and the needs and expectations of consumers.

The introduction of any product, even one so seemingly simple as a different style of hamburger, into a fast-food chain can involve several unexpected, interwoven variables, which need to be assessed. One fast-food chain "recently spent $1 million on thousands of taste tests to develop a better hamburger" (Anon., 1986). This chain experimented with nine different buns, over three dozen different sauces, three types of cuts of lettuce, two different sizes of sliced tomato, not to mention ten colors of four different boxes and some several hundred different names. It was even determined that the order of the condiments was important to consumers. All in all, this new product introduction represented a formidable task in market analysis!

Chapter

8

Product Development for the Food Ingredient Industry

New ingredients and ingredient technology have grown at an amazing pace. To take one sector of the food industry as an example, nowhere is this more apparent than with milk and the many products that can be derived from it, each with its own unique flavor or functional property that it contributes to foods in which it is used. A quarter of a century or so ago, a dairy product tree based on milk would have numbered only a handful of products:

ice cream	evaporated milks	cultured milk
market milk	cream	butter
cheese	dried milks	

Today, milk constituents have been prepared into a wide variety of ingredients (Kirkpatrick and Fenwick, 1987):

- products based on whole milk, such as pasteurized milk, sterilized milk, UHT-flavored milks, and powdered milks;
- products derived from compositionally altered milks such as fat-reduced milks, protein-enriched milks, lactose-reduced milk, sweetened milks, reduced-mineral milk for infant foods;
- milk powder products, with heat stability properties or high dispersibility from modified milk;
- products based on milk fat, ranging from the common cream, butter, and anhydrous milk fat to compositionally modified milk fats with better spreadability or altered fatty acid composition to fractionated milk fat with controlled and defined melting ranges;
- products based on the proteins found in milk: whole protein coprecipitates, components of proteins such as casein rennet, whey protein derivatives

(whey protein concentrates), lactalbumin, milk proteins combined with nonmilk proteins, and modified protein fractions;
- whole cheeses, cheese powders, sprinkle-on cheeses, reduced-fat cheeses, processed cheeses, modified cheeses;
- products based on lactose, where its low sweetness can be utilized, and on the other end of the sweetness scale, its enzymatic conversion to glucose and galactose to produce sweeter products.

As well, there are biologically active materials that can be obtained from milk.

These dairy ingredients find uses in dietetic foods (McDermott, 1987), in meat and poultry products as calcium-reduced binding and emulsifying agents (van den Hoven, 1987), in confectionery products (Campbell and Pavlasek, 1987), and in bakery products (Cocup and Sanderson, 1987).

Other products, such as plant materials, are being similarly purified, fractionated, and blended to produce fiber ingredients with unique emulsifying, stabilizing, or textural properties. Functionally important antioxidants have been derived from herbs and spices. Underexploited plants and fish are finding new uses as new ingredients. And this is to mention only a few.

I. THE ENVIRONMENT OF THE FOOD INGREDIENT DEVELOPER

The same device, i.e., keeping a distinction between "customer" (one who buys a product to use in another product which will be itself marketed) and "consumer" (one who buys to consume) will be maintained, as it was in the previous chapter.

New product development for the ingredient industry presents some very interesting differences from, as well as some similarities to, development in the retail food market and in the food service market. Ingredient developers provide products which some food manufacturer will use in a product for some consumer. In this product, the ingredient developers' products are seldom noted by the general public. Some exceptions here are those diet products containing artificial sweeteners, fat substitutes, or fiber, where the ingredient may be named but not identified by brand.

On the other hand, many ingredient manufacturers have, in addition, a profitable retail market niche for food ingredients, for example baking powders, flavors, colors, various types of flour, meat and vegetable hydrolysates, etc. These markets are subject to all the pressures of the retail marketplace.

Similarities of product development in the food ingredient industry to that in the food service industry are very close, indeed startling. The similarities are threefold:

1. Ingredient development tends to be reactive or crisis-oriented development. That is, the need or demand originates primarily with customers (food manufacturers) who want some ingredient. One need only remember the search for noncaloric sweeteners after the banning of saccharin. In effect, ingredient developers are problem solving, reacting to customers' needs.

2. A corollary of the preceding, product concepts for ingredients are obtained by a thorough knowledge and awareness of the customers' markets. In other words, ingredient concepts originate with customers. This awareness of customers' needs substitutes for the focus group of retail food product development. Ingredient developers, then, must put that concept into a precise concept statement.

3. Ingredient developers serve two masters: the immediate customer, who purchases ingredients for use in a product, and the succeeding customers down the chain, who use that enhanced product and recognize its unique properties. That is, development is directed toward a product (the ingredient) that will satisfy the needs of a customer (a food processor), who in turn will use this ingredient in the manufacture of a product with added-value for some other user. Each customer in the chain puts in added value; each customer has different needs, which must be realized.

There may be many levels of intermediate customers in this chain. Each takes some ingredient, modifies it in some manner or blends it with other ingredients and passes this new modified ingredient with enhanced properties on to other customers down the chain for further treatment. For example, a malting company sells malt to a brewer. The brewer sells the spent malt to a bakery ingredient manufacturing company. After suitable modifications, this modified malt, which now may bear no physical resemblance to the spent malt, is sold to a bakery, where it is used in a baked finished product, which is sold to consumers.

Another example would be the use by flavor extractors of spent seeds and skins from hot pepper sauce manufacturers for extraction of the color and heat principle for sale to confectionery manufacturers and pharmaceutical companies for their products.

The similarity of this to the situation in the food service industry should be apparent. In the food service industry, a manufacturer of a dry soup base, for example, is the end of such a chain. The soup base manufacturer receives ingredients from several sources and blends or alters these to suit the needs of the food service outlet for a multipurpose soup base.

There is one major difference, a difference of degree perhaps, rather than a difference of substance. Development of products for the food service market, whether they be a soup base or a finished product such as a new entree item, is more focused towards the customer/consumer, that is, providing a service. This has meant intense competition, and developers of products for the food service market have opted to concentrate on more specific targeting of their development projects toward satisfying customer/consumer needs.

This is not quite the situation in food ingredient development. Here, the efforts of many food ingredient manufacturers are still largely concentrated on developing properties of ingredients rather than satisfying the needs of their customers who, as in the food service industry, service the needs of consumers.

Ingredient developers are caught up in technology. When they feel they have accomplished something, then go out and broadcast to their potential customers, "Look what I can do!" This is another example of Little Jack Horner research, "... he stuck in his thumb and pulled out a plum and said what a good boy am I." They should be saying to customers, "Look what you can do, and here's how we will help you."

My point can be made clearer if one considers fat replacers. Ingredient manufacturers have shown great skill and versatility in making fat-like imitations from so many different substances at such a rapid rate that the casual observer is given the impression that there is a contest to see how many different substances can be developed into fat replacers. Here developers have produced products that look like oil, have the mouthful of oil, but do not have either the taste of oil or the nutritive properties of oil. They make finished products that are inferior with respect to flavor and texture. Emphasis has been on technology when it should be on satisfying the needs of customers at a price customers are willing to pay.

To use an analogy, a shotgun approach has been used, and the shot used has been one producing a wide scatter. The hunter has attempted to hit everything within the target zone, with the result that nothing has been satisfactorily targeted. A narrow scatter shot, to continue the analogy, might have been more effective in focusing on fewer targets but with more telling effect.

There is one very obvious point of difference with which the food ingredient developer contends that developers in the food service industry or in the food retail market do not face. That is the likelihood of government intervention in the form of safety testing and approvals for the many novel ingredients being developed from new nonfood sources or from genetically altered animals and microorganisms. Indeed, the whole question of calling such ingredients "natural" or even that such products will be termed "environmentally friendly" will be challenged, not only by governments but by consumers as well. Righelato (1987) put this succinctly:

> Regulations exist primarily to protect the consumer, but they are necessarily concerned with existing products and hence serve to maintain the status quo. In doing so they protect the existing producer, who, in fact, probably helped frame the regulations. The ease of introduction of new products in the face of existing regulations depends very much on whether the bureaucracy takes a supportive or adversarial stance.

II. FOCUSING ON THE CUSTOMER/CONSUMER

All product development, whether in retail markets, in food service markets, or in food ingredient markets, should get its inspiration and direction from the customer/consumer, who has some perceived need that can be profitably satisfied. Customer/consumer needs should be the focus. The challenge to ingredient developers is to ask themselves, "How can my ingredient enhance the high quality characteristics of my customer's product, and how can I improve that enhancing ability in that food system?"

The technological ability of an ingredient developer to transform some raw material, be it rice bran or micronized protein, in order to simulate the properties of fats, for example, is unimpressive unless its use has been fine-tuned in some desirable added-value product that the customer can sell to consumers. To repeat, the "look-what-I-can-do" thinking must be replaced by "look-how-this-provides-added-value-character-to-your-product". In the development of ingredients, suppliers cannot be content to characterize the properties of their ingredients and leave it to their customers to try them out. Ingredients must be tailored to satisfy customers' needs and to complement the quality and stability designed into the customers' finished product.

Performance of an ingredient in an artificial food matrix cannot supplant the ingredient's performance in the food systems used by the customer.

A. Customer Research

Customer/consumer research in the industrial ingredient business sector is heavily dependent on the ingredient company's technical sales force. It is their ability to articulate the needs and desires of customers back to technologists in the research and development department that is the company's strength. It is also their primary route to customer research.

Focus groups made up of an ingredient company's clients are clearly out of the question. Most of these customers will be actively competing with one another in the marketplace and will hardly be likely to sit down together to discuss products. Likewise, questionnaires for use in individual, mail, or telephone surveys to gather information about customers are not likely to be successful. These are, after all, intrusions into what may be confidential areas. Application of the Delphi technique of surveying company executives can be of general help to ingredient developers (see Chapter 4).

Demographic and psychographic data about industrial customers are simply nonexistent. There are trade, business, or commodity associations, such as the American Association of Meat Processors, the American Spice Trade Association, the Chocolate Manufacturers Association of the U.S.A., the Milk Industry Foundation, and the International Ice Cream Association, where some very general information can be obtained. At exhibitions and conferences that many of these associations sponsor, some excellent contacts can be developed.

Many ingredient manufacturers have a very high profile in these associations. *Prepared Foods* issues an annual index of trade associations and exhibits (Anon., 1991), as do several other national and foreign food trade magazines.

Once contacts are made, the only sure avenues to customer research are personal interviews. That is, sales personnel from ingredient manufacturers must work hand-in-hand with their customer's technical staff to communicate the problems encountered by the customer. Customer research is very much a hands-on business in ingredient development. Each customer is different, and each customer's product is different. One "blanket" ingredient may not satisfy all the various requirements of industrial users.

Ingredients must provide their industrial customers with finished products that have distinct points of difference. Ingredient users cannot always rely on "off-the-shelf" ingredients to obtain the point of distinction that is desired for their products. Each customer of the ingredient supplier then is unique. The distinctiveness of an ingredient must belong to that customer and that customer alone. Ingredient suppliers do not sell ingredients so much as they sell services.

Flavor houses have developed this art of focusing on the needs of the customer to a high degree. They work closely with customers to produce any flavor sensation their customers want. They can blend from natural flavors, create flavors imitative of natural ones, or create unique flavors not found in nature for customers. Ingredients and the technical service supporting the use of the ingredients are designed to satisfy the customer's needs.

The food ingredient industry is then much more a service industry, a distinctive service industry. The ingredient developer's goal should be to provide a quality service with a product distinctively designed to meet the perceived needs of the added-value manufacturer. To develop ingredients, the supplier must work with the customer backwards, so to speak, saying, "What does this manufacturer need, and how can I satisfy that need competitively?" Development is directed to these needs and their gratification.

It is no longer sufficient for an ingredient developer to supply a family of ingredients, each with slightly varying properties, and then go to manufacturers with a series of samples, saying, "Try these. One of them should work." Since each customer's added-value product is unique, it is very unusual that an off-the-shelf ingredient would be totally suitable. Again, there is an analogy with the food service industry; suppliers to the food service industry must adapt their products to conditions in the kitchen, to the skills of the customer's labor force, and to the uniqueness of the style or type of the food outlet itself and its consumers. Developers of food ingredients must be willing to adapt their processes and products to satisfying the equivalent requirements of their customers.

The developer of ingredients for the food industry requires an intimate knowledge of the customer's problems, market needs, and business. Without this knowledge, development could fail, or market niches could remain hidden.

Development of ingredients for the retail food ingredient business is not unlike development for other retail food products. Standard consumer researching techniques provide the necessary information that permits selected targeting of consumers or the development of specific niche markets. Heavy promotion through cooking schools and cooking demonstrations in schools, church basements, carnivals, and agricultural fairs, plus recipe booklets and free samplings, usually accompany retail sales. Feedback from these promotional tools provides its own consumer research information.

B. Consumer Research

Is there an opportunity for consumer research in the food ingredient field? The answer is an ambiguous "yes and no".

For the "yes" side, ingredient developers can make themselves aware of the consumers' activities in the marketplace. That is, they should know about the health concerns of the consumer: i.e., low calorie, low fat, high fiber, no cholesterol, and no salt are in favor. Nutritious and dietary foods, once relegated to the slow-moving section of the supermarket, are now mainstream and prominently displayed. "Green" is in, and food manufacturers are attempting to draw attention to the "green changes" in their products. Natural ingredients in the list of ingredients on the label are perceived as giving a clean image.

So, "yes", ingredient manufacturers can research consumers, determine consumer trends, and fabricate ingredients incorporating these elements into their products. In a sense, they leapfrog their customers, the industrial users.

This leapfrogging can be used for what Lee (1991) describes as "proactive product development". In "pro-active product development", ingredient manufacturers bring to fruition product concepts that, if adopted as products in the marketplace by consumers, would result in heavy usage of the manufacturer's newly developed ingredient. That is, the ingredient makes possible products (for which, it is understood, there is a marketplace need) that food manufacturers could neither produce previously without this ingredient nor produce at a reasonable cost nor produce at an acceptable quality.

Textured vegetable proteins, surimi-based products, and the mycoprotein product, Quorn, developed by Rank Hovis McDougall are typical examples of such ingredients. All find wide use as analogues in various engineered consumer products (surimi: Duxbury, 1987 and Brooker and Nordstrom, 1987; Quorn: Godfrey, 1988; Best, 1989; and Bond, 1992). A consumer desire for new texture is identified, which manufacturers can satiate with new products using the textures these new ingredients provide.

For the "no" side, "pro-active" product development to create a need for or ability to manufacture a hitherto unrealized product is expensive. To create an ingredient capable of doing this is doubly difficult and expensive. Ingredient developers must have a very accurate and intimate knowledge of consumers.

Impossible? No. Difficult? Yes. Such products generally require extensive development resources (surimi, an exception, had a long history of development and application) and require the education of consumers to accept the product or to learn how to adapt it to local food traditions. Then, ingredient developers must convince food manufacturers of this opportunity and rely upon manufacturers to develop, market, and promote the products. A further complication is the need to establish the safety of new ingredients.

C. What Criteria for Screening?

These are politically and economically unsettled times. There are attempts at rationalizing trade through free trade agreements in several areas of the world. Changes will come. There will, no doubt, be a similar rationalization of food legislation between all the countries signatory to these agreements. Ingredient developers are forced to be aware of such legislative activities or to anticipate possible changes that harmonization of legislation may bring. Any changes in legislation could affect the acceptability of ingredients, and changes in trade barriers may affect the costs and the availability of ingredients or the raw materials they are made from. Needless research time and money spent on developing nonpermitted ingredients must be avoided. Again, time spent working with customers while problem-solving their product development must not be wasted experimenting with ingredients that do not conform to local, national, or international laws or to religious customs respecting labeling, packaging, or product standards.

The sometimes heavy hand of legislated standards and regulations is much more a factor in ingredient development than in the development of retail food products or products for the food service industry. There is currently no rationalization of food legislation, although efforts in that direction are being made, slowly, tediously, and ponderously, through the various Codex Alimentarius committees.

Ingredient manufacturers distribute their products more widely than do other food manufacturers. Ingredients, generally speaking, carry no ethnic, cultural, or nationalistic biases. Consequently, ingredient suppliers usually have large export markets. Two criteria for screening, unrelated to the ingredient itself, arise from this opportunity:

1. Can ingredient suppliers adequately service foreign markets and provide the technical sales and marketing required to support these markets?
2. Do ingredient manufacturers understand the foreign market and its local customs sufficiently well to identify potential users and establish a rapport (recognizing the difficulties that a different language and business customs might impose) with these users?

Obvious solutions are the use of local agents familiar with local conditions in the foreign markets or the establishment of satellite operations in the foreign country. Both have shortcomings. The use of agents interposes one more hierarchical level between customer and supplier through which communication must be filtered. Ingredient manufacturers end up relying on others, their agents, for market information. Satellite operations, unless they have all the facilities of the parent company, do not obviate the need to send samples from the foreign customers back and forth for experimentation. This causes delays and inconveniences for customers. Both avenues represent added costs: agents want fees, and satellite facilities are costly to maintain with the double teaming they require.

Financial criteria for new ingredient development have different time horizons. Return on investment can be accepted over a longer period of time, measured more in years rather than in months. That is, ingredient developers do not expect a payback in 3 to 6 months, as might be expected in the retail food market. Ingredients do not require the same level of the expenses of advertising and other promotional gimmickry that retail food products do.

> An ingredient which has been well researched and developed should be filling a market need and will sell itself to some extent, whereas it is often necessary to create a market for a new consumer product by intensive advertising (Lee, 1991).

Nevertheless, ingredients must be profitable. Unfortunately, there is no easy way to evaluate financially new ingredient development. Financial criteria discussed under retail food product development could be applied in assessing the financial success of ingredient development.

A novel criterion is introduced, similar to that encountered in the food service industry. Costs of ingredient development must be balanced against the customer's financial criteria. After all, the cost of the ingredient in relation to its usage level must not force the user to price that user's products out of competitive ranges. High ingredient costs can force users to seek alternatives.

D. The Ultimate Criterion: Test Marketing

New ingredient launches are usually heralded by announcements in trade magazines and technical journals or demonstrated in the carnival atmosphere of food ingredient trade fairs. The common routine for most ingredient suppliers at a trade show is to hand out free samples of a foodstuff in which the new ingredient has been used. Admittedly, this is primarily a gesture of goodwill and generosity. It is frequently a hazardous thing to do: products are presented under less than ideal conditions and are not examples of "best foot forward" perceptions of what ingredients can do.

There is no test market for ingredients. That is, ingredient manufacturers do not select a geographic area representative of targeted customers (perhaps potential users?) and proceed to launch new ingredients supported by advertising and promotions.

What is more likely to happen is that ingredient manufacturers, after extensive business and customer research, will target potential high-volume users of their newly developed product. They will conduct individual presentations, with demonstrations to show each candidate the values that will come from using their new ingredients. These should be carefully rehearsed and well researched beforehand.

Potential customers evaluate an ingredient on the basis of advantages that accrue from its use:

- Does its use reduce costs per unit?
- Does its use reduce labor costs, simplify production, or somehow increase production efficiencies?
- Does its use increase the high quality shelf life of the product? Is there a satisfactory quality/cost ratio that justifies using the new ingredient?
- Are there added nutritional or safety benefits for products using the ingredient that are valued by the consumer?
- Can the consumer perceive and value the advantages that using the ingredient gives the product?
- Does this ingredient permit the development of products previously impossible to manufacture? Will the use permit the manufacture of a more "environmentally friendly" product, avoid the use of chemical preservatives, and provide a clean label?

There must be affirmative answers to these questions if the developer of ingredients is to know success with the newly introduced ingredient.

III. THE FUTURE FOR INGREDIENT DEVELOPMENT

New ingredients and biotechnology appear to go together. These days, one cannot read of one topic without the other somehow coming up. Best and O'Donnell (1992) review a number of future new ingredients, many of which are of a biotechnological origin:

- bacteriocins of microbial origin,
- transgenically altered milk production to produce, for example, natural preservatives, pharmaceuticals, altered milk-fat fatty acid profile, or to remove a milk allergen producing off-flavors in UHT milk,
- genetic manipulation of poultry to improve texture and water-binding properties of the meat.

They also temper the promise of the new ingredients with several cautions. The first, which has been noted before, is the probable intervention of government in the permission to use genetically or biotechnologically derived ingredients. The second is the impact such novel techniques may have upon the economics of, for example, the dairy industry. Already, this particular industry faces two conflicting government interventionist policies. On the one hand, in many countries, including the U.S., the industry is encouraged to produce milk with a high milkfat content. On the other hand, government health policies exhort consumers to eat less fat. Finally, such new ingredients will bring enormous pressures on ingredient manufacturers to produce, separate, and purify the products to the degree that the industry will demand.

The desire of consumers for natural products and products manufactured using natural ingredients will pressure ingredient suppliers to explore their use and the use of minimally processed derivatives from these natural ingredients. Manufacturers of food products for consumers will also pressure ingredient suppliers for products that will give their labels an ingredient list that appears less chemical and more natural. Natural ingredients will be "in", but whether this will also mean transgenically modified ingredients or biotechnologically derived ingredients will also be considered "natural" and therefore "in" is a moot point. With the consumers' fears of big science so prevalent, a massive educational program will be required to allay these anxieties. Ingredient manufacturers must use caution.

Less highly processed stable foods, which consumers seem to want, will require the use of natural preservatives and antioxidants, in concert with minimal stabilizing systems. Attractive foods will require natural colors. This intense interest in natural colors, natural preservatives, natural antioxidants, indeed anything natural that can replace "chemicals" is attested to by the abundance of research papers and reviews of literature in these areas: natural colors (Engel, 1979; Francis, 1981; Gabriel, 1989; Shi et al., 1992a; Shi et al., 1992b), antimicrobial agents (Beuchat and Golden, 1989; Zaika et al., 1983; Zaika and Kissinger, 1981; Shelef et al., 1980; Daeschel, 1989; Baxter et al., 1983; Barnby-smith, 1992), and antioxidants, (Kläui, 1973; Pokorný, 1991).

These natural foods are being researched intensively by ingredient suppliers to identify the active components. The first companies to bring acceptable ingredients to market will find their efforts rewarded.

The need for natural ingredients with functional properties, such as preservation, thickening, emulsifying, coloring, and so on, as well as with nutritional/pharmaceutical properties, will grow. A major caution to the use of all new ingredients in general, and ingredients of biotechnological origins in particular, will be their acceptance by consumers. A major requirement for that acceptance is that the use of any ingredient in a food satisfy a consumer's need for whatever that ingredient contributed to the food. It is not enough merely to satisfy a customer/manufacturer's need.

9

Organizing for New Product Development

In an earlier chapter, creativity and innovation were described by their dictionary definitions, and explanations of their role in new product development were detailed. A fuller discussion is necessary here to introduce some of the challenges of organizing for new product development.

Interpretations and definitions of what constitutes creativity and innovation abound in the literature. For example, H. J. Thamhain (Dziezak, 1990) describes innovation as

> a process of applying technology in a new way to a specific product, service or process for the purpose of improving the item or developing something new.

Bradbury and co-authors (1972) put forward a more precise definition of innovation, which consequently narrows its meaning:

> a process which originates with the recognition that an opportunity or a threat exists and which is concluded when a practicable solution to the problem posed by the threat has been adopted or a practical means of grasping the opportunity has been realized.

The latter continue with definitions for discovery: "finding or uncovering new knowledge", and, leading directly from this definition, invention is "discovery which is perceived to possess utility".

Whatever the definition or interpretation that one wishes to apply, creativity, innovation, discovery, or invention requires harnessing and encouragement. Companies interested in new products usually attempt to do this by organizing their staff in some meaningful way to direct their personnel's activities and to manage their physical resources to foster creativity and innovation. The question, "Meaningful to whom and for what?" is begged. This will be addressed later.

Organization implies planned systems of predictable activities, all of which activities are coordinated one with another in a controlled fashion. Ultimately, these systems should be interfaced with the other systems that, together, make companies function effectively. If the unexpected happens in this network of systems, the organization of the company should be so structured that remedial activities swing into action to control the unexpected event.

This company organization can be likened to the organization of the human body, which is very highly structured. When the unpredictable happens to the human body, for example an invasion by some virus, defense mechanisms, as represented by the body's immune system, come into play. A very elaborate system of activities are coordinated to combat the viral invader. That is organization.

On the other hand, innovation and creativity are generally recognized as involving activities that are unpredictable, uncoordinated, uncontrolled, uncontrollable, and certainly, to some degree, unplanned. This being so, does not a paradoxical situation arise whereby, if organization is imposed, creativity and innovation are stifled? Actually, less coordination, less control, less planning, and generally less bureaucracy do not lead to more creativity or innovation but to chaos and randomness (Aram, 1973). For example, if a development group cannot be productive, innovative, creative, or inventive within time and budget constraints imposed by senior management who are responsible for the business goals of the company, there is no expectation or likelihood that it will be more effective in these endeavors with no limits provided.

Organizing for new product development seems, then, almost to be a statement of contradiction. In reality, there is no contradiction here. Innovation and creativity cannot flourish in a bureaucracy, with all the pejorative connotations of this word. On the other hand, some organization is required if for no other reason than to keep the activities of technologists, engineers, market researchers, and production personnel under fiscal control and provided with support resources.

I. DEFINING RESEARCH

The word "research" means different things to different people. For instance, to the person on the street, the word conjures up images of complex laboratory set-ups with white-lab-coat-frocked scientists, who are far removed from normal mundane life and activities. This lay person would deny ever conducting research. Yet, this same person will examine brochures about cars, visit several car showrooms, hold discussions with and question numerous car salespersons, and bargain with various financial institutions for the most advantageous payment terms. He would never dream of calling any of these activities research.

There are two broad classifications of research: basic, or fundamental, research, which may also be called pure research, and applied research. Fundamental research is very loosely described as research for the sake of knowledge without thought of commercial exploitation. The descriptive terms for the two categories are by no means clearcut. For example, "basic research" has also been used to describe research for which there was a possibility of exploitation (Gibbons et al., 1970). Applied research, also referred to by Gibbons and colleagues (1970) as mission-oriented research, is research directed to some specific goal.

The following classification may remove, or perhaps add to, the confusion:

- interest-for-interest's-sake research, which can best be described as research with no foreseeable application. It is the dilettante's research, just to satisfy curiosity (Gibbons et al., 1970). Muller (1980), in a very interesting paper decrying bureaucracy's stifling of innovation by pettifogging funding policies, might term this "seed" research, i.e. research time and effort to follow up ideas. There are no time constraints in this research;
- basic (pure) research, which (no matter how much the academics might argue to the contrary) is always undertaken with some expectation of an application in the future, if for no other reason than to get a research grant. Nevertheless, Muller (1980) quotes Wernher von Braun as having said, "Basic research is what I'm doing when I don't know what I am doing." If time constraints exist, and they usually do, they are measured in years rather than months;
- goal-directed research, which, as its name implies, is research directed to some very specific mission, the application of which is quite apparent. The application need is at most a few months. This would be described by Gibbons et al. (1970) as mission-oriented research.

Readers must accept for themselves the meanings that apply within their environment.

Most new product development would fall into the goal-directed category. This is especially true for small companies, which cannot afford any other type of research save that directed to protect and expand their profitability. Large companies may separate their research and their development departments. Even here, with the exception of multinational food companies, research is confined generally to "basic" research, that is, with some expectation of future exploitation. This research is usually directed to objectives to be accomplished in 3 to 5 years. Developmental programs are focused on goal-directed research aimed at very specific, short-term, food product objectives.

An interesting recent development is the pooling of research resources. Large corporations pool their resources with government/university consortia, whereby cooperative ventures dedicated to longer-ranged projects of research can be undertaken.

Interest-for-interest's-sake research is rarely undertaken knowingly by food companies. It has never, to my knowledge, been engaged in for new food product development. Some explanation of that statement is required. Research with no foreseeable application may be undertaken by food companies to accumulate knowledge or to gain experience in some area of science. While there is no foreseeable application, nonetheless, companies may have ulterior motives (not based on the science or its outcome, necessarily) in doing the work. For example, support of graduate students in some esoteric field of food science qualifies as interest-for-interest's-sake research. However, there is a good likelihood that those graduate students may be hired by the sponsoring companies. Was this research with no foreseeable value? Sponsoring companies had ample opportunity to evaluate the candidates.

Research with no foreseeable application (interest-for-interest's-sake research) may be undertaken within a company. It is rare that management knows it is undertaken. Such research has picturesquely been called "underground research and development" (Aram, 1973). Aram, in a study of a company involved in research and coincidentally in innovation, noted that informal networks developed within the organization. It was through these informal systems that innovative research occurred. Aram found "... the cross-departmental informal organization ... had the connotation of an activity that was disguised, if not almost illicit. Part of its attraction and its effectiveness seemed not to be managed." Aram cites two individuals from one particular underground group, one from new product sales and the other from product engineering, whose group was responsible for ten patent applications.

II. CONSTRAINTS TO INNOVATION

A. The Corporate Entity

The reluctance of food companies, and especially large, multinational corporations, to engage heavily in basic research is easily understood. It is expensive. While any corporate entity would welcome a major technical breakthrough that could enhance its competitive edge, this desire, however, must be tempered by corporate financial goals, by shareholders' desires to make an annual profit, and by the need to stay within the financial constraints of an annual budget. Management, including corporate management, has a time horizon rarely fixed more than a year or two ahead. Graduates of most of the business schools have been drilled to believe that profit is the name of the game, and profit is rarely viewed as anything more than short-term profit. Short-term profit has no interest in long-term research. It is regrettable because, as Dean (1974) put it succinctly, looking to short-term profit "... is like looking for the leak in the bottom of my canoe as I drift toward the unseen waterfall."

Marketing and production departments have time horizons set closely within the annual budgetary plans of the corporation. As a consequence, these departments are usually smiled upon favorably by senior management.

Scientists and engineers, on the other hand, look to a much more distant time horizon. Senior management, looking at bottom lines on a quarterly and yearly basis, is much more apt to cut the long-term research projects when it appears that budgets may be exceeded.

Risk capital is expensive. Simply put: why take risks with large expenditures of money on innovative and creative research if the rewards cannot be assured or may not justify the expenses incurred? Only a very small proportion of research ideas ever leads to the development of some innovative product. Even patents are equally unlikely to be financially rewarding. Indeed, only roughly 2% of all patents survive their full life.

Furthermore, within any large, technologically successful organization, there may not be a strong incentive to embark on any heavy program of new product development. If management believes that the company has a proven superiority in technology and skills over its competitors to provide products that have a seemingly never-ending global acceptance, there is little impetus to engage in a heavy program of innovative product development. The philosophy, "If it ain't broke, don't fix it", prevails. Companies in the fast-food business have used the philosophy that if one product can fit their global franchise requirements, they do not need to change their strategy. (However, many of them now seem to be broadening their menus, and local franchisees are experimenting with local cuisines.)

No less a personage than Akio Morita, chair of Sony and inventor of that "innovative marvel" the Walkman® (trademark of the Sony Corporation), was reported as saying (Geake and Coghlan, 1992) that companies place too much emphasis on basic research to their own detriment. Their reliance on basic research prevents their being competitive. The Walkman® did not contain any new technology. Its secret was in new packaging and marketing. Morita saw certain qualities in various new developments and put these together with the perception of a consumer need. This is innovation.

Another corporate concern about long-term research is the inordinately long lapse of time between the actual discovery or invention and the development of some innovative application in the form of a new product. One estimate puts this at roughly 11 years (Bradbury et al., 1972). Disappointingly the company reaping the benefits of the new technology frequently is not the company that made the original discovery. These observations do not encourage companies to engage in long-term research.

B. Communication Problems

Communication problems between people, between departments in the same company, or between manufacturing plants within the same company division are difficult to deal with at any time. In new food product development, they can be particularly disruptive. Product development teams need to work closely as cohesive units. Unfortunately, communication problems frequently lead to conflicts of personalities. Which came first, the people problem or the communication problem between them, between their departments, or between their manufacturing plants, is a moot point.

It is a common complaint that research and development departments in large, multinational companies with several different manufacturing plants are too isolated from the mainstream activities of the manufacturing companies that they are supposed to serve. A gulf develops between the research and development department and other corporate activities, such as marketing, production, and finance, as well as between the regional plants. Each of the other company departments is as concerned, albeit each in its own fashion, with new product development. Nevertheless, this seeming isolation causes a communication gulf to widen.

This gulf also springs up between technical management and the top management of other corporate activities. What causes this division is not easy to pinpoint. One can generalize by stating rather simplistically that scientists march to a different drummer than do marketing or production or financial people. But what does this mean?

The misconception that the management of scientists and technologists is somehow "different" was spawned by the rapid growth of the concept that research and development was a potentially valuable corporate activity. Mismanagement of technical resources resulting from this misconception may perhaps have contributed to the gulf between the scientists and other corporate functions.

Communication problems are exacerbated in multiplant companies. In these companies, communication between plants (lateral communication) or even between the technical staff within these plants can be poor or even nonexistent. For example, I was informed in one plant of a multiplant company that a particular manufacturing problem had been successfully resolved several months previously. This same problem was unresolved at a sister plant (making the same product!) not 600 miles distant where I visited a week later. Lateral communication between the several plants of this corporate giant was poor at passing information to or even sharing information with other plants. The result was duplication of effort.

Large multinationals have tended to centralize their research and development resources. The reasoning is fairly easy to follow. By incorporating all the expensive research and development equipment in one facility, together with all the pilot plant equipment and libraries to support the technical staff, there will be great economies of money, no duplication of facilities, and better communications, yes? No, not always. This has frequently produced corporate ivory towers of research and development divorced from the technical and developmental problems of regional plants, which has only exacerbated communication problems.

For example, I have found some of the following communication problems in large, multinational, multiplant companies with centralized research departments:

1. Laboratory personnel of a branch plant, when being advised of the upcoming visit of the vice president of research of the international parent company, were routinely warned not to discuss with him any projects in which they were engaged should they be asked. Staff were required to clear their laboratory work benches of all apparatus.
2. In a company where the head office research and development laboratories and technical library rivaled that of many small universities, inquiries by laboratory personnel situated at several branch plants to this center were discouraged by the local plant managers for fear that such contacts would result in "them" meddling in, or taking over, research projects in the branch operations.
3. A multinational company conducted a new product development program at its corporate research headquarters in Europe on a confectionery product and test marketed the product there. The product was destined for the North American market. The product failed when introduced in North America.
4. In a similar situation, a fruit juice developed and test-marketed in Europe at the international corporate research and development headquarters of another company was packaged in material neither approved for nor available in the North American market. It, too, was meant for the North American market.

These are unconscionable and inexcusable breakdowns in communication. They depict the worst examples of the polarization of effort and lack of communication that can occur in the management of the research and development in large, far-flung, corporate ivory tower research centers.

Technical communication or technology transfer, i.e., the dissemination of useful technical developments within the company, requires different managerial skills than does the management of technical people; the skills are quite distinct. To manage technology transfer within a company requires someone with communication skills, not necessarily someone who can prepare a soundly designed experiment or write a good technical report. To manage technologists and scientists requires the ability to encourage and inspire people, to protect them from bureaucratic intervention, and to challenge them.

Managing research is the function of managers of research establishments. Management of the transfer of technology, a quite different matter, should not be relegated to laboratory managers (or supervisors or directors of laboratories or project leaders) but to those individuals with skills to communicate.

Not understanding this distinction leads directly to another form of communication breakdown: the transfer of technology from the research and development resource to centers within the company that could utilize the information. The transfer fails because of the NIH syndrome discussed earlier; this is variously interpreted as "not-invented-here" or "not-interested-here"; either interpretation fits. When people have not had an opportunity to be part of the development and have not been encouraged to see how this development might assist the objectives assigned by management, a sudden attack of NIH may and usually does occur.

Technical people frequently have a problem communicating with nontechnical people. Nowhere is this more apparent than when those nontechnical people are from the marketing department. Marketing people live in a world of optimism, chutzpah, hyperbole, and persuasion, where sooner rather than later is more appropriate. Technical people, by contrast, prefer a world of logical methodology, where organized skepticism is the rule. Technical people live in a world of verifiable facts and want to keep perfecting and testing, in other words, seeking protection and solace behind irrefutable data; this is much to the annoyance of marketing personnel. Technologists are devil's advocates, doubting Thomases. They are inveterate tinkerers, who want to keep on perfecting, never wanting to let their pet project go. In this, they, like others, can become very attached to their pet project.

A common complaint of marketing personnel is that research and development people are intractable and inflexible and do not or cannot respond to the rapidly changing environment in the marketplace. This latter charge deservedly earns them the criticism from marketing people that technical personnel are against everything and lack imagination. Technologists do not understand that the introduction of a new product must be timed precisely and that speed is essential. So say the marketers.

Technical people complain about time. Marketing does not give them time: time to research and develop the project, time to test all variables, and time to retest and retest.

Marketing and technical people speak different languages and use different measuring tools in their trades. Each is skeptical of the merits of the other's tools of the trade. The vagueness ("airy-fairyness", as I heard it put) of terms used in quantitative scaling techniques in consumer research and concept testing disturbs the technical person used to logical methodology and verified data with statistically significant results.

This language issue can be a very real one. For example, I well recall using the words "rheological properties" in a product development review meeting to describe the flow properties of a freeze-stable icing system. This engendered hoots of laughter from the marketing personnel amid pleas to speak English. Yet, they felt no discomfort about dropping such terms as "perceptual mapping" or "nonmetric multidimensional scaling analysis" during the same meeting.

Production and marketing personnel gripe with one another all the time, whether new product development is involved or not. Marketing people complain of poor deliveries with too many defectives, resulting in consumer complaints. Production departments, of course, dig in with demands for better forecasting schedules, especially when marketing departments, in their wisdom, embark on promotions requiring extra product or special packaging materials, a situation that leaves production staff scrambling to obtain supplies or having to employ extra work crews. Production personnel live in an ordered world ruled by scheduling of labor, of supplies, and of produce. Any disruption to this order affects their bottom line. New product development, especially plant runs, disrupts order and may disrupt production. Any disruption could affect bonuses. Innovation, if it is not properly communicated, can be strongly resented, unless in the communicating, the advantages have been described fully.

Tensions arise and are quite normal during the push for new product development. They arise between all segments of the new product development team, between marketing, production, and technical personnel. What one does not expect and does not want are problems at the interface between each of these groups that result in the breakdown of communication.

This, then, is the environment of problems and constraints in which new product development must be managed productively and efficiently. Whether the company is large or small, the same problems exist. They differ only in degree in proportion to the company's size.

C. Personnel Issues

Job security concerns all personnel. One's personal security within a company will certainly influence one's productivity. If productivity is measured by one's innovativeness and creativity, one's contribution to product development will suffer.

Failure rates in new product development are horrendous (see earlier). By the odds, any new product development venture is expected statistically to fail. This is the environment in which all on the product development team live. Production personnel are the least vulnerable members of the team when there is a new product failure, but marketing and technical personnel stand on the front line facing the odds of failure.

In the event of a new product failure, corporate management must be a just and forgiving management. Management must allow for failure in such a high risk enterprise.

To repeat, when there is a product failure, it must be carefully analyzed to determine what factors were incorrect or what caused the project to fail. This analysis must be conducted constructively; it is not a witch hunt but a learning experience from which all can benefit. On the other hand, errors must be rooted out and corrected. Weaknesses in the development process must be strengthened. Reassignment or retraining of staff may be necessary, which management must handle positively, encouraging their staff's development.

In the same manner, in the event of a successful product launch, just as much is to be learned by an in-depth analysis of why and how success was attained. A keen understanding and knowledge of the strengths of the total development process will be invaluable for future projects. A secondary benefit derives from this analysis when management can suitably reward the achievement of the team members. This secondary benefit can do much for fostering innovation by providing a sense of security and appreciation of one's effort.

Close on the heels of the preceding is the need for management in the organizing for innovation to develop and encourage young scientists and engineers for the future of the company's growth requirements. The rewarding of achievement, as well as the learning by analysis of both new product development successes and failures, will promote the growth of new skills within the company.

Management must itself accept some blame if innovation has been constrained. If management is unable to defuse the conflicts one naturally expects between such disparate groups as marketing, production, and technical personnel and cannot oil the frictions discussed above, then too much pressure will be put on personnel, who then will be unable to produce innovatively. Innovation will die. Morale will suffer. Management must be able to unify the new product development team and encourage them to work as a cohesive force.

Another human problem must be introduced here. This is more a feature of the multiplant, multinational organization. Much product development work, especially that which is involved with leading-edge technology, is multidisciplinary work. That is, the services of specialists, who may be employed by the same parent company but work at separate locations or who may be consulting academics at distant universities or research institutes, can all be required in a project. Members of a team may never meet except electronically or may meet only rarely face-to-face to exchange ideas. As a result, a gulf develops between team members. This remoteness can be a dampener to the team spirit of the product development group.

Companies with such development projects must make sure that all the distant members contributing to the project are assembled together frequently enough to exchange ideas personally rather than impersonally. This improves communication among the team. An added benefit can be the increased productivity created as members of the team interact by bouncing ideas off one another.

III. ORGANIZING: FOR WHOM? FOR WHAT?

There is no shortage of management information, replete with charts for the organization of research and development. With solid lines and dotted lines describing lines of either authority or communication, they can be very impressive. A reader wanting these can refer to several excellent papers, which are still very pertinent despite their age:

- Mardon et al. (1970) discuss at some length the problems of administrating technical departments of multiplant companies.

- The role of the technical manager is discussed from a very human perspective by Head (1971), who sees this manager with only two resources, people and their skills and material resources (equipment and laboratory facilities) and as one who "... inherits a situation, good, bad or indifferent." How to make the best of the good, bad, or indifferent situation is discussed.
- Aram (1973) describes those informal networks (the undergrounds) that evolve in companies for research and development and supports their encouragement rather than any attempt at their formal management.

The above sources describe in both specific and general terms the management and the structure of technical organizations within companies. Head (1971), on organizational charts, has the following delightful comment:

> Devotees of the organisation chart take it up as a personal challenge to find a box for all of them. Ask what they all do and you will probably be submerged in a torrent of peculiar terminology about 'line control finance wise' and 'inter- and intra-functional communication channels'. Initially, one thing only will be clear — the appalling debasement of the English language.

In considering the organization of research and development for product development, one must get beyond charts. Yet, many companies live by their organizational charts and dote on the dotted and solid lines linking the various boxes. Head would most surely have been a devotee of Aram's unstructured underground.

In the opening to this chapter, it was stated that companies try to organize their staff in a meaningful way for product development purposes. Two questions emerge: meaningful for whom? and meaningful for what?

Meaningful for whom? Organization for new product development is meaningful largely for managers in order to develop a cohesive team of diverse talents and to motivate and direct these toward the creation of some specified product required for the company's business plans. "The future manager will become steadily more active in catalyzing the participation process among his subordinates: equally, he will expect, in increasing measure, to participate with his own masters" (Head, 1971). Organization is not, then, for lines of command necessarily nor merely for lines of communication but for lines of participation. Participation is a strong element in Aram's underground research groups, which form around those who can contribute and participate.

Meaningful for what? Organization is meaningful largely to facilitate communication; communication implies participation, which in turn implies that those who participate also contribute. Communication is not meant solely for keeping the lines open between the multidisciplinary segments that make up the new product development team but also open up and down the pyramidal structures, with their vertical lines of communication that develop in large corporations and which can isolate one group from another. Lateral communication is necessary as well.

Hierarchical organization for product development should stop short at the product development manager. That is, motivating people in directions the company desires and providing lateral communication to encourage teamwork, especially between the technical and marketing departments, are functions best left to the product development manager.

Large companies attempt to control people and the skills they possess through a formalized hierarchy of inter- or multidisciplinary management teams (portfolios, as one company terms them). Small companies have more informally structured organizations, in which, as in large companies, personalities can dominate. Today's new product development manager must be able to ease the project through the various departments involved and smooth the way for the development of a strong, cohesive team spirit.

Managers of product development have basically two resources, physical plant and people (Head, 1971). Managers must learn to harness physical as well as human resources. Plant is inert, immutable. People are the most promising in terms of creativity and innovation. Innovative people will use physical plant effectively and efficiently. The touch of the manager must be deft: too much control, too much pressure, can stifle creativity and innovation. On the other hand, too little control provides no certainty that innovative product development will ever result. Organization is necessary to a degree; otherwise chaos would rule in the company. However, organization is not the equivalent of management; organization cannot manage innovation and creativity. Management can foster and encourage these.

Fluidity of movement is essential within any new product development group. As part of the technology transfer, technical people must be prepared to be transferred with a product as it matures from the laboratory bench through engineering and production to marketing. In this manner, language differences soon disappear. All members of the team talk the same language or at least are understood by the other members of the team; this is the important point. Communication laterally between team members and vertically within participating departments must be encouraged. Managers must have communication skills to sell both technology as well as the innovative skills of the members of the group to others vertically and laterally within the system. The members of the new product development team come from different disciplines within the company, which have different sets of values or interpretations of company objectives. The manager must make this a cohesive group. By moving technologists with products that they developed, the professionalism characteristic of many technical people is meshed with the commercial and business interests of the company. A greater understanding of the contribution of all results.

To foster growth, managers of new product development will be successful communicators first and facilitators second.

Chapter

10

Back to the Future

What will the future bring for new food product development? The statistics presented by Friedman (1990) and Kantor (1991) show that new food product introductions have been increasing year after year for nearly 30 years. During the latter half of the 1980s, the increase in introductions has been meteoric (see Chapter 1). Indeed, by extrapolation of Friedman's data, one could anticipate something approaching 30,000 new products by the year 2000. Thus, one is tempted to predict that there will be more and more new products year after year.

There will be no end to the number of new food product introductions if the statistical trend is to be believed. This prediction of a staggering 25,000 to 30,000 new product introductions each year after the year 2000 begs two questions:

1. Is such a simple extrapolation justified? Are there influences that have not been factored in to the projections that might indicate that there will be a slackening in the numbers of new food product introductions in the foreseeable future?
2. Which new products will these be that will be introduced in the future? More of what?

These questions will plague new product developers, and there does not seem to be any easy answers for them. A simple extrapolation is not justified for reasons that will be elaborated in this chapter. There may already be a slackening of introductions.

What the new products of the future will be is the question marketers are asking as they peer into their crystal balls. If they knew, they would dearly love to proceed with the research and development required for them now to be ahead of the competition.

Any attempt to predict precisely the nature of new products that the future may bring tomorrow or the next day or the next is doomed to failure, because the outcome depends on the next enactment of food legislation, the next scientific or technological breakthrough, the next press conference on the findings regarding the toxicological, nutritive, or curative property of some food component, or the next ecological or environmental finding regarding food processes or agricultural practices. These are all unknowns.

I. PAST DREAMS AND PROMISES

Hindsight is a wonderful sense. One should learn from hindsight. As I read with hindsight the predictions made many years ago by others concerning what we would have been eating today, I should learn not to make predictions. That is certainly true when attempting to predict the future of new food product development. If only there were some rhyme or reason, some element of predictability, to food product development.

The science and technology on which food product development are based is both unpredictable and, as yet, underdeveloped. Food technology outstrips its underpinnings of food science. The result is that technology has had to rely on trial-and-error and craftsmanship in equipment design, product formulation, preservative technology, and ingredient technology.

> ... man had always been concerned with the technology of food, because he needed food to survive and that his understanding of the basic principles underlying the art, or craft, had always been slower than its development. Coppock (1978) (see also Taylor, 1969)

Without the science, applying technology for new product development will always be costly in time and money.

To even think of consumers collectively introduces a false premise. Consumers are volatile. The "traditional" family no longer exists, and with it have gone many traditional food habits. The proportion of elderly in the population has grown, so that it has now become a recognizable pressure group within society. The sciences associated with understanding the consumer and what influences the consumer's choices are developing rapidly and changing our concept of the consumer. The "average consumer" no longer exists if, indeed, there ever was one. The average consumer entity has been shattered into many fragments, each representing some unique market niche. Clearer identification of consumers certainly will help developers in formulating new products and marketers in promoting them.

Politics and government are taking an increasingly active role in food and agricultural policies. The results are an increasing body of legislation regulating food and increasing support for farm, agricultural, and seafood policies, a growing body of health and nutrition policies, and a growing tendency to liberalization of trade policies. All of these can be unsettling for prognosticators in the food business.

Reviewing earlier predictions of what was to have happened in food and agricultural areas can be both sobering and quite educational. Some forecasts have come to fruition: some have proven to be pie-in-the-sky dreaming, and some may yet come to be. Why were some predictions dead on and others not so successful? An examination of some of these may be helpful.

Whitehead (1976) wrote enthusiastically in 1976 of the changes that might be seen in 1999:

- Meals might include algae-fed oysters from a sea farm. An entree of mock (spun soy protein analogue) chicken would follow, with a mixed vegetable casserole in a base of single-cell protein, topped with a cheese analogue. Dessert would be cake baked using triticale flour.
- The superfarm might be laid out in long narrow strips spanned by moving bridges, on which sits the farmer who, from his position on this bridge, can program all activities, including cultivation, seeding, irrigation, weeding, and harvesting of most crops.
- "Farm communities of 1999 will be totally integrated with atomic energy plants which will supply electricity and heat …" and "… farmers wearing cybernetic equipment may be able to control their machinery by thinking about what they want it to do."
- Aquaculture farms, particularly salmon and catfish farming, will grow in importance. So, too, will lobster and oyster farming gain greater prominence.
- The soft drink industry will develop a market for syrups and powders that are carbonated at home.
- Insects, cattle manure, poultry droppings, and municipal sewage will provide the basis for protein for animal feed.

Some of the above are technically feasible. However, for social, practical, or economic reasons, they have not been undertaken.

In 1977, after a year-long study, the American Society of Agricultural Engineers made (Anon., 1977) some very long-range forecasts up to the year 2076. These mainly predicted changes in farming practices, but they did include some food predictions:

- By 2076, most meals (75%) will be prepared at large food service commissaries and not in the home.
- Fruits and vegetables will be grown in solar-heated greenhouses near or in urban centers. Fresh fruits and vegetables will be delivered to consumers daily.
- Poultry, as well as other livestock, will be raised in environmentally controlled high-rise buildings as part of urban areas, with feeding and waste management automatically controlled. Organic wastes will be extracted for their energy.

- Harvesting and in-field processing of field crops will result in the separation of the edible material into protein, carbohydrates, fats, and other useful food ingredients. Cellulosic and ligninic material in the waste will be converted into plastics.

There are more than 80 years to go for these forecasts to be properly judged. Already, the majority of eating occasions are taken away from the home. This prediction has almost been fulfilled within the first 20 years. But will this still be true in 2076? The impact of the personal computer and networking facilities, as well as improvements in telecommunications, has made the concept of the home-office a reality. What will be its impact on meals-away-from-home?

All the above predicted events have been accomplished now or have been demonstrated to be possible although perhaps not practical now. This begs the question of why there is this extraordinary lag between discovery and application.

Where the projected time frame for predictions is closer to the present day, predictions are usually more accurate. The vagaries of technology, the government, and the consumer have not contrived to distort the data on which these forecasts were based. A study, released by Frost and Sullivan, entitled *New Food and Beverage Products Market* (reported in Anon., 1980a) gave the following predictions for the 1980s:

- The use of analogues in new food products will expand, especially for microwave-designed products.
- The use of whey-based ingredients, high-fructose corn syrups, encapsulated flavors, and savory flavors will expand.
- Food service menus will broaden their menu selection. Salad bars will be more prominent in these establishments and offer a wider selection.
- Government influence will continue to grow in nutritional labeling and dietary claims.
- Nutritionally positioned products will displace naturally positioned products. Nutrition will be the "turn-on" for consumers in the future.
- The over-65's are one of the fastest growing segments of the population. The geriatric food market will present a challenge to food product developers and marketers in the future.
- Ethnic foods will grow and cause segmentation of the market.

These predictions, covering a shorter forecasting period, describe the 1980s and early part of the 1990s very well. The fast-food restaurants are certainly attempting to go up-scale with white-tablecloth seating and broader menu selection; expanded deli salad bars are to be found in fast-food restaurants and supermarkets. A recent wire service article (*The Gazette, Montreal*, August 16, 1992) reported that McDonald's and Burger King have both experimented with special dinner menus.

Nutritional products are prominently positioned now, but one might argue whether the natural foods (no additives) or minimally processed foods have been displaced to any extent. In addition, some believe the bloom is fading from nutritionally positioned products.

Ryval (1981) made the following predictions but wisely put no time element on them:

- Meat consumption, particularly red meat, will drop and with it animal fat consumption because the replacement foods, such as lentils, contain little fat. In addition, more fresh fruit and vegetables will be consumed, as well as more minimally processed foods.
- Food selection will broaden, with a greater acceptance of ethnic foods such as hummus, curry, tofu, and tempeh.
- Seafood consumption will rise. There will be a greater reliance on aquaculture and fish farming. Underutilized species will receive greater recognition as food in products such as fish sticks, fish sausage, and surimi.
- A more diverse variety of plant protein sources will be used to complement the traditional soybean, peanut, and cereal grains.
- Analogues will continue to gain favor, not necessarily as total replacements but rather as extenders of other products.
- Waste recovery techniques will undergo intensive research to develop new feed sources by fermentation with microorganisms or to extract valuable food ingredients from the waste.
- The retort pouch "spells the end of the tin can" because of the energy savings the pouch offers, plus the opportunity for improved quality.
- Sterilized milk will move milk out of the refrigerator cabinet and onto the store shelves.

Again, in the short time since Ryval (1981) collated these prognostications from experts, one can see that some have come to pass, while others have sputtered out or continue to arouse nothing but yawns in consumers. What has made the difference?

Aquaculture has proven to be modestly successful but not the panacea it was once thought it would be in making seafood readily available. There are still disease problems to be overcome, and on the Pacific west coast, there are pollution problems caused by the aquaculture industry itself, which are threatening its existence.

Analogues from various protein sources enjoy a good success in many products where they can be blended or incorporated structurally with their natural counterpart. Single-cell protein, particularly yeast protein, has some success as a base for food ingredients. Triticale flour has never become popular, and I have only seen it in some health food stores.

Insects as food or feed require a great deal more consumer education than most companies will risk undertaking. Cultural taboos will have to disappear before consumers eat insects or before consumer activists knowingly allow animals to be fed insect-derived protein or manure-derived feed. Similarly, consumers have not fully accepted atomic power, especially in their backyards or farmyards, as the case may be.

There are curbs both to an unlimited or unchecked growth in new food products and to the acceptance by consumers of new food products. Simply put:

1. New food product development is expensive. The sciences related to consumer research and food and nutrition are not fully understood. Because of this, there is much trial and error in development, and this costs money.
2. Legislation pertaining to food, its labeling, packaging, advertising, safety, and its manufacturing will continue to be a thorn in the side of new food product developers, as will governmental support policies and trade affiliations.
3. Consumers are fickle.

Through the history of food technology, one has been told that freezing as a process was going to replace canning (thermal processing of food). Glass bottles were to replace metal cans and become the container of the future. Then the retortable flexible pouch and its cognate, the semirigid container (thin profile containers), were to be the packaging of the future. Tetrabrik would replace the glass bottle and plastic would replace everything, if one were to pay attention to all the hyperbole in the advertisements. None of these predictions came to pass. Foods processed in glass and metal are on the shelves side-by-side with plastic-packed foods and Tetrabrik-packed foods. North Americans have never really adapted to thin-profile containers, despite their huge success in Japan and modest success in Europe.

Who could have anticipated the greening trend that initially swept Europe and then came to North America? Proponents and users of plastic packaging rapidly started to back-track as antipollution legislation took effect. Biodegradable was in; plastic was out. Glass bottles were looked on with more favor because they were potentially reuseable. They could be easily reconverted or, when crushed, could be used in paving or landfill. Packaging companies clearly had a problem in attempting to find ways to make their products recyclable.

II. FACTORS SHAPING FUTURE NEW PRODUCT DEVELOPMENT

It is not possible to predict with certainty which new products will emerge in the future as successes. One can, nevertheless, certainly forecast events and technology that will shape the nature of these new products. These can be classified into three broad areas:

1. social concerns including a growing environmental concern, the so-called greening revolution (distinct from the "green revolution" of the 1950s); better informed and educated consumers; consumerism and consumer mistrust of bigness, whether it be business, science, or government; and changing food consumption habits;
2. scientific and technological advances in such diverse fields as biotechnology, human nutrition and disease, and farming practices (e.g., animal husbandry and organic farming);
3. factors in the marketplace that will encompass competition, government legislation, changing trade patterns as nations enter into economic alliances and "free trade" blocs, the resulting social and political upheavals these alliances may bring, and the global market for food and agricultural products that will result.

A. Social Concerns

The "greening" movement means many things to many people. For all, this greening movement is, somehow, related to environmental concerns, which term is itself an all-embracing if not downright confusing one. Some see involved in the movement attitudes toward third world nations, agricultural practices respecting organic farming, factory farming, animal rights, and health issues. To others, it relates also to marketing and marketing policies. Subtly, the greening of the food industry includes issues of food processing and waste management, utilizing raw material sources that do not inflict cruelty to animals, using pure and natural food, recycling water, and conserving energy usage, as well as truth-in-labeling, adopting a corporate green policy, eschewing overpackaging, using recyclable packaging, and promoting good corporate citizenship. Greening becomes a mushy confusion of issues to which, nevertheless, companies should pay attention. Some of these issues will be discussed in detail.

1. Pollution and Environmental Concerns

Pollution and the environment are major concerns especially among people in the developed world. They are an emerging concern among indigenous peoples in the less-developed world. This concern resulted in a major international conference, "Earth Summit", which was held in Rio de Janeiro, Brazil, under the auspices of the United Nations Conference on Environment and Development, June 1–12, 1992. No longer is action being demanded solely by a vocal minority of "do-gooders" but by governments and nations, which see that something must be done to reduce pollution. At one time, waste of any sort could conveniently be dumped somewhere in oceans, lakes, rivers, or landfill sites, and forgotten. Medical wastes have washed up on the eastern seaboard beaches of the United States and have been clandestinely dumped in foreign countries (Patel, 1992). Cities and nations can no longer afford land space for garbage disposal, and today one hears of boats, trains, and trucks loaded with waste destined to travel forever, it would seem, as country after country refuses them entry. Disposal of waste has become a significant cost for food companies, who must pay to have waste hauled away and then pay a tipping fee at a land disposal site.

Waste production must be managed. That will require that less waste be produced in the first place. Products or processes that produce excessive amounts of waste will be rejected, not only because of their inefficiencies in the utilization of a resource but also for fear that environmentalists may protest against the companies involved in producing the waste. Or, companies will have to divert some of their technological skills and resources to improving processes and products to make them less wasteful and to upgrading the waste into useful byproducts.

The packaging industry has been most affected by the pollution concern because of its very high visibility. Food containers are designed to give optimum protection to the product from harvest areas to the processing plant and thence throughout the stable life cycle of the finished product. Akre (1991) rather glibly states:

> In the developing world up to 50% of available food stuff is lost to rats, mould, rotting, etc. because of poor packaging and distribution. In the developed economies however no more than 2% of available food is lost, thanks to packaging.

The container is the most obvious feature of processed foods. When these empty containers are strewn along highways, dumped into streams, or caught in vivid photographs of small animals caught by the neck in plastic ring connectors, environmental activists become very concerned and vocal. Governments listen.

Germany will require that 64% of many types of packages be recycled by mid-1995. By agreement, the Dutch government and its packaging industry will ban landfill use, effectively committing the industry to an extensive recycling program by the year 2000 (Akre, 1991).

Product developers are, therefore, compelled by public and governmental pressure to either use recyclable or reusable packaging or biodegradable material for packaging. Both alternatives present developers with a dilemma. Recycling or reusing requires a collection system. This is an added cost; also, some argue that it is not energy efficient. Biodegradable packaging material is more expensive and is not readily biodegradable in most landfill sites (Lingle, 1990). The use of degradable packaging materials raises several problems, not the least of which is the reaction of regulatory officials to the uses of the degradable materials in contact with foods. A similar challenge may be put to the edible films that are emerging.

At present, new product developers should consider recycling as the better alternative to degradable packaging.

As a side note to this pollution problem, a recent news broadcast (April, 1993) reported that some municipalities would fine residents who put out recyclable material or compostable waste for refuse collection. Residents must compost their biodegradable refuse.

2. Less Processed, More Natural Foods

Indirectly, two things have led to a desire by many consumers for less processed foods and more natural foods, these environmental concerns, as well as a general reluctance to accept highly processed foods, for which the consumer understands "ersatz" and "somehow not good". Lee (1989) refers to "food neophobes", i.e., those people who consider new food technologies and food additives and ingredients as untried, artificial, and hazardous. Busch (1991), when discussing the consumers' concerns about biotechnology, puts it this way:

> They desire foods that have been prepared in traditional ways, that contain few or no additives, that are "natural", and that are made neither from transgenic plant or animals nor via new fermentation techniques.

This distrust of processing and high technology may be challenged and dismissed as irrational, but nevertheless, it is there and it is growing. (It should be noted, however, that consumers espouse biotechnology in all its subdisciplines in matters of health and vanquishing disease.) The interest of consumers in minimally processed chilled foods may reflect this desire for less processing and more naturalness.

Also, the demand for organically grown products, even organically grown added-value products, has increased. Jolly and his colleagues (1989) reviewed some startling historical statistics:

- A 1965 survey conducted in Pennsylvania showed that only 15% of respondents held a great deal of concern for pesticides, while a similar survey conducted in the same state in 1984 found 78.7% to hold a concern about the use of pesticides.
- Of respondents, 94% in 1965 felt there had been adequate inspection of food purchased at retail, but this figure had dropped to 48.9% by 1984.

These figures indicate a growing lack of confidence in the safety of the food supply.

Jolly et al. (1989) found that 57% of all respondents in California judged organic foods to be better than nonorganic foods, and 35% considered them to be no different. Obviously there is a perception that organic produce has advantages. These advantages were listed in order of importance as no pesticide/herbicides, no artificial fertilizer, no growth regulators, and residue-free.

Campbell (1991) reviewed two Canadian surveys of consumers' concerns about the food supply. When asked what was their most serious food concerns, most respondents answered pesticides (26%), pollution (24%) was next, and nutritional concerns next (15%). When a multiple response was permitted, the answers were food poisoning (72%) and pollution and pesticides (tied at 69%). Food additives rated low as a perceived hazard.

Whether scientists think pesticides are a hazard or not, other people — nonscientists, the consumers — do.

"Big Firms Get High on Organic Farming", read headlines in the respected newspaper The Wall Street Journal (Nazario, 1989) in 1989. Many farmers and large companies with extensive farm holdings (Sunkist Growers, Inc. and Castle & Cooke, Inc.) have joined the movement to natural or organic farming. They find there is a demand for their organically grown products. Many realize that organic farming can be as profitable as so-called chemical farming. As Nazario (1989) wrote, "… case studies have found that yields and profits can be just as high on an organic farm as on a nonorganic farm."

It is worthwhile for any food technologist to visit a natural food store to examine the products and the people purchasing them. They should even compare organic products with competitive nonorganic products. A complete line of fresh produce is available at elevated prices, but so deep is the distrust of nonorganically grown produce that people are buying despite the price differential. The demand is so great that large farming companies have moved into organic farming with their more efficient agricultural practices, and prices for fresh produce are falling. Irrational it may or may not be, but there is a strong demand for organically grown produce and the added-value products made from them. The organic food business is no longer "… a counterculture business run by flaky hippies" (Nazario, 1989). McPhee (1992) reported that organic food introductions have seen an increase of 400% since 1986, and the organic beverage category has increased 1450%. The organic food industry is not without its problems. As McPhee noted, big companies are still testing the market cautiously, well aware that any movement into this new market could have repercussions with their regular lines.

The Institute of Food Technologists (IFT) published a Scientific Perspective (Newsome, 1990), in which an objective assessment of organically grown foods was presented. It was pointed out that the claim that organically grown foods are healthy is scientifically baseless. IFT highlighted problems with organically grown foods, citing the *Listeria* outbreak in cabbages caused by the use of sheep manure (natural fertilizer usage). Newsome concluded in this *Scientific Perspective* that:

• Organically grown foods were not superior with respect to quality, safety, or nutrition to conventionally grown foods.

- They are more expensive.
- They provide less variety of product for consumers.
- Diets based only on organically produced foods may "... present the risk of possible loss of balance and variety ...."

This is absolutely correct. But,

- if consumers perceive there to be an advantage,
- if consumers are willing to alter shopping habits to go to specialty stores to purchase organically grown products,
- if consumers are willing to pay more for organically grown foods,

then one must admit that there is a certain qualitative "superiority". The consumer is right. Hauck (1992) recently described a ranching and meat operation in the U.S. that provides organic beef and lamb with sales of $25 million in 1991, projected to $50 million in 1994. There is clearly a market for organic foods for consumers who believe them to be superior.

The consumer is becoming better informed, better educated, and more vociferous. The result is a very strong consumerism movement. Food scientists and technologists in their ivory towers are being challenged. Their science and technology have apparently outstripped their communicative skills. Perhaps this has been a factor in a consumer mistrust, not only of science, but also of big business and, equally, big government (for example, see O'Neill, 1992). Communication between consumers, on the one hand, and science and technology interests, big business, and big government is not well developed (Lee, 1989; Busch, 1991). Until communication has been improved and the vitriolic rhetoric that has developed is curbed, consumerism and all its manifestations may be a prominent consideration of new food product developers in the future.

3. Changing Food Habits

Consumers are becoming more interested generally in food and cooking. Proof of their interest is the burgeoning numbers of cookbooks describing a wide gamut of ethnic cuisines, the growing popularity of cooking schools, and the popularity of television cooking shows. Consumers are doing more cooking from scratch; that is, consumers have gone back to using traditional recipes for home cooking, especially for special festive occasions. Experimentation with new cuisines has led to the use of more exotic ingredients. This awakening to foods and cooking has provided food manufacturers with opportunities for new products.

Families eat fewer meals together. This has meant that more meals are eaten away from home. In addition, grazing (snacking) is an established and preferred eating pattern for many people who accept several small meals a day as normal practice. Such a change in eating habits opens up opportunities for finger foods that are tasty and nutritious. Already in response to the grazers, food pushcarts are making their appearance on the streets of many of the larger metropolitan areas. They serve the function of the fast-food restaurant, without the overhead of expensive real estate. Many of these pushcarts sell ethnic foods.

There is a serious side to meal patterns and their influence in daily life. Chrononutrition, an off-shoot of chronobiology, the study of biological rhythms, has emerged as a study of the "time-dependent features of nutrition" (Arendt, 1989). How do time or timing of meals and the consequences of digestion affect food selection or other biological rhythms in the body? The practical application for these studies is the concern for shift workers' or airline crews' safety and efficiency when their daily rhythms are disrupted. The gastrointestinal problems of shift workers can lead to work loss, fatigue, and inattentiveness. Worker safety is endangered.

Foods, particularly snack foods, could be so designed as to provide the proper nutrition, based on the findings of chrononutrition, to workers whose biological rhythms are disrupted by their work schedules.

Ethics will have a greater role in shaping the purchasing habits of future consumers (Wilson, 1992). Guides are available for shoppers that provide information on the companies behind the food brands: for example, *The Ethical Shopper's Guide to Canadian Supermarket Products*, authored by Helson et al. and the staff of EthicScan Canada and published by Broadview Press. These guides describe a company's environmental record, its policy on women's issues, labor relations, and consumer issues in general. There have been similar developments in the investment community, where stock investment plans are devoted entirely to purchase of stocks of companies whose policies are ethical and "green".

The influence of the so-called "alternative formats" for retailing food, the warehouse outlets, direct mail selling of food items, as well as direct sales, selling directly into the home, has not yet been fully assessed as to the changes these will bring to the food marketplace or even whether such alternatives will survive in the future. However, these alternatives will influence the development of new package formats and new products to satisfy the demands of these channels. Conventional retailers will have to combat the impact that alternatives have on consumers and their buying habits.

4. Vegetarianism

Vegetarianism means many things to many people. Some vegetarians eschew red meats and partake of poultry and seafood products. Others avoid all animal foods where there has been a killing but will eat animal products such as milk, dairy products, and eggs. Still others refuse all animal products. The reasons for adopting a vegetarian diet are many. Some do so for religious reasons. Some avoid animal products because of a respect for animal rights and on humanitarian grounds. Some argue health reasons for having a vegetarian diet — they "feel better" — and consider vegetarianism a healthier lifestyle.

Some regard the rearing of animals as a waste of global resources. And some just like vegetarian foods better.

Those concerned about their health can be influenced to become vegetarians when headlines in a respected science magazine state, "Surgeon General Says Get Healthy, Eat Less Meat" (Anon., 1979). These reports are then copied by every science editor of every newspaper chain. "If I eat less meat and get healthy, then if I eat no meat, I'll be even healthier." Such is the thinking of some of the public.

Vegetarianism is growing rapidly. Krizmanic (1992) reported that a recent survey by a prominent marketing research company found that 7% of the American population consider themselves to be vegetarians. At one time, it was impossible to find a restaurant that had any main course vegetarian dish on its menu; now, restaurants may feature several main course vegetarian dishes, and there are restaurants that are exclusively vegetarian. Most, but not all, of these restaurants are ethnic, and their menu items are a source of new product development ideas. As the vegetarian movement establishes itself and as more consumers experiment with the kaleidoscopic assortment of dishes that are possible, the quest for the highly processed, low-fat foods, such as burgers, may be bypassed by consumers for the juicier, spicier, more flavorful lentil or tofu burger with very low fat content, excellent nutrition, and no additives. Already, a St. Louis-based meat product supplier has introduced frozen beef patties that are supposedly over 90% fat-free using a vegetable protein. It could be an excellent marketing ploy — surf-and-turf giving way to earth-and-turf.

B. Impact of Technology

There is a latent fear of technology in the nonscientific community. The cause is irrelevant. The nonscientific community, which includes most people, is either confused by, frightened of, or suspicious of technology, except when it is applied to the pharmaceutical and medical devices fields. There has been a recent controversy over the value of cow's milk for children. The pro's and con's of this argument, i.e., whether the use of cow's milk by young children resulted in diabetes in later life, have been played out in the newspapers, on radio phone-in shows, and on television. Some mothers are afraid to give milk to their children. Experts, medical doctors, nutritionists, and dieticians on both sides of the issue argue for their respective positions. No wonder the public can harbor a suspicion of science and technology, when the proponents of the disciplines cannot agree on the simple issue of feeding children milk.

Another case in point is the issue over aluminum in Alzheimer's disease. There are press releases to counter the association of aluminum with this illness. Contradictory press releases claim that the disclaimers of the influence of aluminum are wrong. Is this how the public, the scientifically and technically illiterate, are going to be educated to the value of science in their lives?

With press releases and press conferences to publicize scientific discoveries, particularly in medical and nutritional matters, followed shortly thereafter by disclaimers, retractions, or statements, such as, "... it is premature to think that ..." or "... much more work needs to be done..."?

1. Big Science: Biotechnology

Biotechnology is a growing field of endeavor, with great promise in the medical field, where already it has been used with success in the treatment of some genetic disorders. This promise to relieve pain and suffering is welcomed by all.

The public is not so welcoming when biotechnological techniques are applied to the food supply, where this branch of science promises to be of even greater value to food processors than it has been in the medical field. Here, suspicion, caution, politics, greed, big business, big science, government, and consumer advocacy groups all seem to cloud the issue of what impact biotechnology could have on agriculture and food production. Meanwhile, consumers are complaining that they do not want furry potatoes or hard, bruise-free tomatoes. There are no perceived consumer advantages.

Bovine somatotropin (BST), derived from biotechnology, has been used to increase milk production in cows. Consumer advocacy groups are suspicious of its long-term use on people consuming the milk products derived from the milk. Small dairy farmers are frightened by the prospect that large dairy factories that can afford to use BST to increase milk production will force them out of business. Their dairy associations lobby government for protection.

Leaf culture techniques now permit scientists to grow tobacco leaf or "natural" vanilla in factory vats situated far from the natural geographic source (see, for example, Knorr et al., 1990). But what will be the impact on tobacco farmers or on the economies of nations such as Madagascar, the Comoro Islands, the Island of Réunion, Tahiti, and Mexico, for whom the vanilla bean is an important commodity? Fear, suspicion, and politics must be overcome before technology, and especially biotechnology, will be a major contributor to new product development. It is not enough for biotechnologists to control the gene that controls senescence in a tomato so that distribution losses can be controlled. Tomatoes are not meant for distributors; they are meant for consumers. Will there be a perceived advantage for the consumer? That is the key issue.

Similarly, breeding and gene selection programs for sturdier banana trees that resist disease better and have straighter bananas that pack better and resist transit damage are fine advantages for growers and distributors. But these are not advantages easily perceived by the consumer, no matter how much they may be appreciated by the grower and shipper. "Will my bananas be different?" laments the consumer.

Again, the introduction of manufactured microorganisms into the environment for nitrogen-fixation or as a cryoprotectant for plants is an issue understood by scientists. There is no advantage to consumers. And it is grist for the mills of the environmentalists, who can readily shake the confidence of the consumer by suggesting the hazards, real or imagined, that such an introduction may bring. Consumers remember the chlorofluorohydrocarbons (CFCs). CFCs were touted as harmless, biologically inert, noncombustible substances. Then, the world, including consumers and scientists, discovered the effect CFCs had on the ozone layer. All of a sudden, something considered harmless when used in large quantities became dangerous.

Harlander (1992) stated the following:

> For some, biotechnology is the key to enhanced productivity and improved quality, nutrition and safety of the food supply .... For others, food biotechnology is something to be feared and carefully regulated, since the potential long term socioeconomic or environmental effects cannot be predicted. If biotechnology is to be used to ensure a safe, abundant and affordable food supply, it must be accepted by the public; therefore it is critical for us to come to grips with the scientific, as well as the social, moral and ethical issues that influence our thinking about the food supply.

A difficult task, indeed. For biotechnology to be accepted in new food products, it must be perceived as being an advantage or improvement for the consumer, as being environmentally safe or neutral *over the long term* and not merely as an advantage for the grower, shipper, distributor, or manufacturer. (For a fuller discussion, see Harlander, 1989.)

But closer to the consumer than any of the foregoing was the article by Molly O'Neill in *The New York Times* for Sunday, June 23, 1992, headlined "Geneticists' latest discovery: public fear of 'Frankenfood'." "Frankenfood" is, of course, genetically altered food, a name coined by Paul Lewis in a letter to the editor of Th*e New York Times*. The newspaper article featured developments at the exhibition of the Institute of Food Technologists (1992), in particular, a frost-resistant tomato incorporating a fish gene. This may very well have advantages for a grower, but where is the advantage for consumers? If the consumer cannot see the need for it — that is, the consumer's need for the modification — or appreciate the advantage the modification gives to the consumer but perceives it as an unnecessary alteration of an accepted product, then no amount of education will convince the consumer to accept this product *at this time*. Already, companies have backed away from some of their genetic research.

A good comparison may be events in the early introduction of irradiation techniques. Irradiation as a food preservation technique was originally touted as being able to prolong the shelf life of foods. This was perceived by consumers as an advantage for the processor, not for consumers themselves (Best, 1989). The consumer would be getting staler product. Is this not logical reasoning on the part of the consumer? Now, the issue of irradiation for food products is very clouded with highly polarized views on all sides of the issue, which only serves to confuse and arouse fears in the consumer. The extent to which apologists for science will have to go in educating the consumer can be seen in a review by Pszczola (1990) describing some of the rhetoric.

2. Factory Farming and Agricultural Practices

The green movement spawned many things in its meteoric growth. There were concerns about animal welfare. The practices of animal husbandry, as carried out by so-called factory farming techniques, were descried. On occasion, there have been violent demonstrations against both food retailers displaying factory-farmed poultry and veal and against farmers employing factory farming techniques. Car bumper stickers proclaim, "You hug an animal called a pet. Why do you eat an animal called dinner?" Changes in the raising of animals have been severely questioned for their humaneness, by both vegetarians and meat-eaters alike. Even traditional practices, such as force feeding to produce *foie gras*, have been declaimed. My purpose in describing these incidents is not to present an argument for vegetarianism nor against factory farming, but merely to suggest that there are social constraints on the technologies and crafts that are applied to food production and processing.

When these agricultural practices are compared, e.g., the feed conversions of raising beef or milk vs. the number of people who could have been supported by that equivalent amount of grain, there is a growing awareness on the part of many people that perhaps the way things have been done are not the best. That is, more people could have been fed using the grains not as animal feed, but as food for people. The question for developers to ponder is whether a growing social conscience will put a curb to certain new food product development in products of animal origin but give, on the other hand, an impetus to the growth of products of plant origin?

Triticale, an early, expensive biotechnologically developed crop, has never gained wide popularity. The rediscovery of old food crops will have an impact on agricultural practices as developers create new products from them to satisfy the consumer. Rediscovery is perhaps not an apt word, except in a North American context, as crops such as millet, quinoa, and amaranth have long been favored in other cultures. Amaranth, in particular, has been described by Brücher (1983) as "the poor man's high-protein cereal". Brücher provides a brief history and description of the crop and information on its cultivation. Teutonico and Knorr (1985) have reviewed the properties, composition, and applications of both the grain amaranth and the vegetable amaranths.

3. Eating One's Way to Health: Nutrition and Pharmafoods

Knowledge of the interrelationships between nutrition and disease has progressed well beyond knowing, in the 1930s, the need for vitamins in the diet, lest scurvy or rickets developed. There is a growing awareness that the foods people eat may play an important role in preventing the onset of cancer (NCI, 1984; USDHHS, 1984; Anon., 1985; Ames, 1983; Cohen, 1987; Maugh,

1982), heart disease (Taylor, 1980), and other diseases (Jenkins, 1980; Berner et al., 1990). Diet may also play some role in how people behave (Wurtman, 1989; Kolata, 1982; Erickson, 1991; Barinaga, 1990).

There are important social and political implications if, by eating foods able to prevent a disease or by using added-value foods enriched with these disease-preventing foods or with the responsible component extracted from these foods, people may avoid certain diseases. Or, if they can reduce their risk of developing certain diseases by eschewing certain foods, then implications for governments are powerful. Social programs depend on these.

For consumers, there is the potential for a healthier lifestyle and a longer life by eating the "right" foods. This means an even greater growth in the number of elderly. This has implications for the work force, retirement policies, pension plans, medical plans, the health care system, agricultural policies, food policies, and companies wanting to capitalize on people's fears.

Already, there is a large and very lucrative market for what has been called mood foods, nutrifoods, pharmafoods, nutriceuticals, or medifoods. These are foods, extracts, and herbal preparations that have some real or imagined effect on people's health, intelligence, sexual prowess, or mood. They are promoted very aggressively in strength, body-building, and fitness magazines and are usually distributed by mail order. With such claims, legislation cannot be far behind. Justifying the claims and establishing the safety of the products will add enormously to the cost of new food product development in this area. Nevertheless, the lure of trying to produce a nutritious and tasty product that protects against one's chances of developing cancer (for example) will be strong for any product developer.

Current evidence clearly suggests that certain diets composed of specific food items are better at reducing the risks of some diseases. For example, some recently published articles are the following:

- Kritchevsky (1991) reviews the effect of garlic on cardiovascular disease and discusses its hypolipidemic, hypotensive, anticoagulant, and fibrinolytic properties. This is certainly a good bet for an enterprising developer to come up with a pharmafood.
- Mills and co-workers (1992) discuss work on the biochemical interactions of peptides derived from the breakdown of foods and the possible influence these may have on food intolerances, work of immense value to consumers suffering food allergies. Two possible spin-offs from this work the authors cite are the screening of foods for precursors causing intolerance and the possibility for developers to remove such precursors from new foods.
- Bioactive peptides derived from milk proteins, e.g., opioid peptides, immunopeptides, and mineral-binding peptides, were reviewed by Meisel and Schlimme (1990). These authors speculate on the role of these derived dietary products as food hormones and "natural" drugs.

- Bifidobacteria have received prominence lately, and bifidus milk is on the market as a treatment for various intestinal ailments. O'Sullivan et al. (1992) review the subject of probiotic bacteria, describe some of the products on the market, and suggest areas for further work.
- Hughes and Hoover (1991) discuss bifidobacteria in dairy products and describe products already on the market shelves in various countries.
- *Aphrodisiacs* is the provocative title of an article by O'Donnell (1992), which reported on the ability of certain foods to affect human reactions; that is, they can act on sensory perception (cf., alcohol or hallucinogenic mushrooms); they can irritate or they can stimulate. These "mood foods" offer opportunities for new products.
- The antioxidant hypothesis regarding cardiovascular disease, in which oxidized cholesterol is the risk factor and not cholesterol, was reviewed by Duthie (1991). In this, insufficient antioxidant intake to prevent the oxidation of cholesterol and associated free-radical activity is the problem. This clearly suggests a tie-in of vitamins E and C, beta-carotene, and selenium, all food nutrients.

As research data accumulate, the evidence for these health benefits will mount, and new food product development will be influenced. People will demand these products. Certainly, there should be a large demand for the "deep-sea protein" food supplement made by a European company and claimed to banish wrinkles (Cremers, 1993). Developers will want to produce these products, and governments will have to provide the legislation and inspection services to protect the consumers with respect to the safety of these products and the truth of the claims made for them.

These findings and the products that could be made based on them, coupled with the promotional activities to support them, would blur the fine line between responsible claims ("good for you") and irresponsible marketing claims. Most certainly, the food industry can expect government intervention in the form of advertising guidelines.

C. Marketplace Influences

There are physical limitations to a continuous and unchecked growth in the number of new food products introduced into the marketplace. There is not enough shelf space for all the new products. Supermarkets themselves cannot continue to grow to accommodate all the new products. Products that do not pay the expected return per running foot of shelf space must be pushed off the shelves. But there are a number of products that, new at one time, have become staples (for example, instant coffee) and cannot be budged off the shelves.

1. Marketplace Changes

Changes in the marketplace are legion. The warehouse outlet is climbing in popularity, as consumers buy in bulk to control food costs. Whether they really control costs or not is immaterial if they believe they do.

The field of communication is expanding at an enormous rate, in particular interactive television. What impact will teleboutique shopping have on the food retail marketplace and ultimately on the presentation of new food products? How does one target a segment of the market that one wishes to reach? Snow (1992) suggests that "instead of broadcasting adverts to the old 'admass' the new buzz word is 'narrowcasting'".

Conceptually, the supermarket is changing. It is still a marketplace all under one roof, but now various departments may be privately owned or leased to specialist tradespeople. The meat department might be owned and operated by a professional butcher; the in-store bakery might be run by bakers; the fruits and vegetables might be run by a knowledgeable greengrocer who cares about the produce, and so on. There is professionalism. One no longer stands bewildered, desperately seeking help from some teenager stocking shelves. In the new supermarket, there are staff knowledgeable about the breads they sell or the sausages they make, and they have samples cooking for one to try. There is someone who can discuss the various coffees and their roasts or the tea blends available. And the consumer is not captive to one butcher, one baker, or one fishmonger. There could be, for example, two or three butchers, bakers, or greengrocers hawking their wares. The whole effect is to produce a carnival-like atmosphere typical of a marketplace where people can eat, meet, socialize, and shop.

2. Legislation and Other Government Intervention: Their Impact

Government intervention will become an increasing burden on the food industry, and especially on food companies heavily committed to new product and new process development. This will be felt in areas as diverse as the following:

- patent protection for genetically altered foods;
- toxicological testing of new ingredients and foods derived through biotechnology, and submission of safety data of these and other new products, such as degradable and edible films, to regulatory authorities;
- development of a nomenclature for biotechnologically derived products;
- labeling regulations and guidelines to define or clarify such concepts as "natural", "nature-identical", "organic", or "minimally processed"; and
- advertising claims for the new ingredients and products with health benefits.

In a good example of the latter, the consumer protection department of the French government has recently cracked down on special diet foods for athletes that contain large amounts of carnitine and for which claims are made that carnitine increases the amount of energy to cells, enhances athletic performance, and reduces the amount of fat in cells (Patel, 1993). In addition, some concern for the safety of products to be derived from biotechnology will be particularly suspect when the microorganisms used in their production are hazardous.

Political and economic factors will influence the growth of new food product development. The awareness of the influence of food and nutrition on human behavior and diseases will force governments to adopt agricultural and food policies that promote healthy food production and reduce the risk of nutritionally related diseases. An attempt by the National Cancer Institute of the U.S. Department of Health and Human Services (1984) to publish a guide to healthy food choices to reduce the risk of cancer caused a furor from some vested interest groups, mainly meat producers, but also some vegetable growers. Yet, such guides to food selection could have many benefits, not the least of which would be to improve the health and productive life expectancy of people and thus reduce health care costs. Health care costs are an important economic factor for any government to consider.

There are new trade patterns emerging that will have an impact on new food product development. Free trade areas are either in place (Canada and the U.S.), awaiting ratification (Canada, the U.S., and Mexico) or being negotiated (the GATT Agreement). If nothing else, these new markets will usher in new competitors with new products to crowd the retailers' shelves. And there will be new consumers to whom to market products. The impact on the movement of raw materials as well as finished added-value goods will be immense, not only within the established zones but also between the trading blocs.

However, the new economic alliances and free trade zones might also bring protectionism, and they will most certainly introduce more food legislation. Free trade zones will open up new markets and both present new opportunities for new products and give new life to old established ones. But there will be upheavals in agricultural marketing policies and farm board subsidization policies. Despite the removal of trade barriers, there will still be impediments to trade, which politics must treat. Vulnerable elements within the agricultural and food manufacturing systems will always clamor for protection, and governments will have to pay them lip service at least. This can only mean that nontariff trade barriers will be an impediment to the free flow of products.

Many experts predict that there will be a global marketplace eventually. Few people would argue with this, but it must not be understood that a global marketplace means global food. Mass food for a global market is most unlikely. The successful food developer will adapt products, their shapes, textures, flavors, and colors to the needs and expectations of consumers in the particular targeted geographical marketplace that the developer wants to penetrate. And the truly innovative developer will be aware that in established markets, the taste and flavor preferences of consumers are being challenged by the greater varieties of ethnic foods available.

No single product will be successful universally in all the marketplaces of the global community. Even the ubiquitous fast-food chains realize this and cater to local tastes in the many countries they have penetrated. And the success of any product in any given marketplace will always be limited by

the social and political upheavals that will plague the world, either in cata-
tyles and eating habits change

to continue to proliferate year
pment process must improve.
money on new product devel-
io must improve, with better
The costs of development must
iinimize the costs and time by
iething as basic as the applica-
s in testing, such as rotatable
: work, reduce the number (and
interpretation of data.
)ut product development. The
that must be recognized. The
:ts may have to be rethought.
essment of profitability by how
nore in terms of whether they
y are recovering their cost of
ore besides.

THE FUTURE

of the future? That is anybody's
)ne can see that there are many
iy to feel daunted at the prospect
:ts. Equally, there are pressures
do anything other than develop
will continue to make new food

characterized, without specify-

iutritionally designed to satisfy
l of consumers. However, the
:alorie, high-fiber, etc. products
analysts. Consumers will want

their final preparation for con-
; seeming oxymoron will be a
:ing systems.

- Finger foods (snacking foods) will become more nutritious and attain a degree of acceptance and sophistication that will blur the distinction between fine dining and "grabbing a bite": for example, the Spanish snack foods, *tapetas*.
- More ethnic foods will become an accepted part of the diet of the North American population, much as Italian and Chinese cuisine have been adopted. Throughout the global marketplace, the variability and diversity of food will make consumers more adventurous.
- Vegetarian main courses (not necessarily vegetarianism) will find wider acceptance, partly through the growth of ethnic foods, partly for health reasons, partly for religious reasons, and partly for social reasons. However, vegetarianism will also become more popular.
- Organically grown foods, and minimally processed added-value foods based on organic foods, will grow in popularity as the general consumer will continue to distrust big government, big business, big science, and big agriculture.
- More meals will be eaten away from the home. Consequently, the food service industry will have to develop new foods to satisfy the increased demand for meals-away-from-home with both variety and health. The work week will decrease, but the need for leisure foods will increase.

The marketplace of the future will change. This will inevitably mean a change in food products. Supermarkets are meeting the challenge of the fast-food chains by opening deli counters where customers can purchase lunches, dinners, and even breakfasts. Street vendors, with their push-cart foods, provide a direct challenge to the fast-food chains, with their high real estate costs.

Opportunities for new food products have been opened up by the revolt of many consumers against bigness, the growing sophistication of consumers' food tastes, and the more cosmopolitan nature of the consumer, but perhaps not in the ways that many food manufacturers would have liked to have seen. Traditional markets have been fragmented, and there is more opportunity for niche marketing — placing new products in markets that are either too small for large companies to fill or are too unprofitable for them to be concerned with. Within these niches are very profitable markets for the right products.

A prime example of niche marketing can be seen in the brewing industry. For years, bigness was in, and there was a limited number of popular beers available. Then, small, local breweries started up. These microbreweries, as they are called in Canada, or craft breweries as they are called in the U.S., kept away from the so-called popular beer tastes and concentrated more on traditional beers. They have become a huge success and have caused some fragmentation of the beer market by targeting beer lovers rather than beer drinkers.

A similar fragmentation has occurred in other markets. Flavored beers, low- or no-alcohol beers, alcohol-free wines, wine coolers, carbonated and still fruit drinks, and flavored natural waters have all snatched part of a market away from established beverage products and created niches that have matured or are maturing into very profitable markets. Snow (1992) considers this move to niche marketing a result of marketing people's uncertainties regarding consumers and consumers' fragmentation of the retail marketplace. In short, the consumers' volatility has fostered new marketing opportunities, which marketers have not quite come to terms with.

In many instances, these new niche products are easier and cheaper to develop and introduce into specialty markets. The developer stands a better chance of getting a return on investment if financial expectations are moderated.

Twenty-five years ago, I wrote a futuristic paper about man's foods in the year 2000 for a contest. Empty office buildings in the heart of the city would serve as the farms of the future. Hydroponic culture would take place in the upper stories of the buildings, and products would pass directly down to lower floors, where further processing would take place. At the bottom, the finished added-value product would be sold or transported away. Waste would be converted along with city waste into animal feed and pumped into other buildings housing confined animals. The process would be repeated. Animals would be fed in the upper stories and passed to the lower floors as they mature. When mature, the animals would be slaughtered and further processed. Byproducts of animal raising (eggs, milk, dairy products) would be processed in stages as they pass to the lower floors, where they would be sold or shipped to other destinations.

I also predicted synthetic foods from leaf proteins and other vegetable protein sources, spun into new textures. The technologies are all there. All the above are possible, but most have only met with modest economic success, if any success at all.

What went wrong? The major factor in all new food product development, and in the basic and applied research that leads up to their development, is the changing consumer profile. My predictions did not factor in consumers and their needs; I satisfied only my needs as a technologist. Food manufacturers must satisfy the needs of consumers.

The above predictions and those that formed the retrospective at the start of this chapter satisfied the needs of technocrats, not the needs of consumers. The developers, including the researchers at their laboratory benches working at the so-called cutting edge of science, must realize that for successful new product development they must satisfy needs of consumers. Sticking one's thumb into a Christmas pie of esoteric research, pulling out some technical plum, and saying, "What a good boy am I" will not be good enough for tomorrow's consumers. Consumers will not beat a path to products simply because they are technological marvels. They will beat a path, however, to products that satisfy their needs.

REFERENCES

Adams, J. P., Peterson, W. R., and Otwell, W. S., Processing of seafood in institutional-sized retort pouches, *Food Technol.*, 37, 123, Apr. 1983.

AIC/CIFST, AIC/CIFST Joint Statement on Food Irradiation, Ottawa, Canada, 1989.

Akre, E., Green politics and industry, *Eur. Food Drink Rev.*, 5, Winter 1991.

Ames, B. N., Dietary carcinogens and degenerative diseases, *Science*, 221, 1256, 1983.

Amoriggi, G., The marvellous mango bar, *Ceres*, 24, 25, Jul./Aug. 1992.

Anderson, J. R., Boyle, C. F., and Reiser, B. J., Intelligent tutoring systems, *Science*, 228, 456, 1985.

Andrieu, J., Stamatopoulos, A., and Zafiropoulos, M., Equation for fitting desorption isotherms of durum wheat pasta, *J. Food Technol.*, 20, 651, 1985.

Ang, J. F. and Miller, W. B., The case for cellulose powder, *Cereal Foods World*, 36, 562, 1991.

Anon., R & D lame ducks, *Chem. Ind.*, 913, 1971.

Anon., Underexploited Tropical Plants with Promising Economic Value, Report of an Ad Hoc Panel of the Advisory Committee on Technology Innovation, National Academy of Sciences, Washington, D.C., 1975.

Anon., The future as seen by agricultural engineers, *Food Eng.*, 49, 120, Sept. 1977.

Anon., Surgeon General says get healthy, eat less meat, *Science*, 205, 1112, 1979.

Anon., Analogs, ethnic and geriatric products offer top growth potential; "natural" to wane, *Food Prod. Dev.*, 16, 52, Mar. 1980a.

Anon., Food and Nutrient Intakes of Individuals in 1 Day in the United States, Spring 1977, Nationwide Food Consumption Survey 1977–78, Prelim. Rep. No.2, U.S. Dept. of Agriculture, Washington, D.C., Sept. 1980b.

Anon., Irradiation for fruits & vegetables, *Food Eng.*, 53, 152, Dec. 1981.

Anon., Nutrition and cancer prevention: a guide to food choices, draft publication, U.S. Dept. of Health and Human Services, Washington, D.C., 1984a.

Anon., Cancer prevention research: chemoprevention and diet, *Backgrounder*, National Cancer Institute/Office of Cancer Communications, Dec. 1984b.

Anon., *Cancer Prevention Research Summary — Nutrition*, Publication No. 85–2616, National Cancer Institute, reprinted Mar. 1985.

Anon., Introducing the hamburger, *Montreal Gazette*, Sept. 16, 1986.

Anon., The creative approach, *R. Bank Lett.*, 69, No. 2, 1988a.

Anon., Showcase: fiber ingredients, *Prep. Foods*, 157, 151, Jul. 1988b.

Anon., Fatal reaction, *Can. Consumer*, 18, 6, Jul. 1988c.

Anon., Modified Atmosphere Packaging: A. An Extended Shelf Life Packaging Technology, B. Investment Decisions, C. The Consumer Perspective, Report Series, Food Development Division, Agriculture Canada, Ottawa, 1990a.

Anon., Natural oxidation inhibitor breathes shelf life into meats, *Prep. Foods*, 159, 71, No. 6, 1990b.

Anon., Psyllium stabilizer: label friendly and functional, *Prep. Foods*, 159, 127, Aug. 1990c.

Anon., Food Industry Source Book™, *Prep. Foods*, 160, No. 13, 1991.

Anthony, S., Clearing the air about MAP, *Prep. Foods*, 158, 176, No. 9, 1989.

Aram, J. D., Innovation via the R & D underground, *Res. Manage.*, 24, Nov. 1973.

Archer, D. L., The true impact of foodborne infections, *Food Technol.*, 42, 53, Jul. 1988.

Archer, D. L., The need for flexibility in HACCP, *Food Technol.*, 44, 174, May 1990.

Arendt, J., Regulating the body's internal clock, *Food Technol. Int. Eur.*, 25, 1989.

Argote, L. and Epple, D., Learning curves in manufacturing, *Science*, 247, 920, 1990.

Babayan, V. K. and Rosenau, J. R., Medium-chain-triglyceride cheese, *Food Technol.*, 45, 111, Feb. 1991.

Babic, I., Hilbert, G., Nguyen-The, C., and Guirard, J., The yeast flora of stored ready-to-use carrots and their role in spoilage, *Int. J. Food Sci. Technol.*, 27, 473, 1992.

Bailey, C., Scaling up in R & D for production, *Food Technol. Int. Eur.*, 115, 1988.

Banks, G. and Collison, R., Food waste in catering, *Inst. Food Sci. Technol. Proc.*, 14, 181, 1981.

Barinaga, M., Amino acids: how much excitement is too much?, *Science*, 242, 20, 1990.

Barnby-Smith, F. M., Bacteriocins: applications in food preservation, *Trends Food Sci. Technol.*, 3, 132, 1992.

Bastos, D. H. M., Domenech, C. H., and Arêas, J. A. G., Optimization of extrusion cooking of lung proteins by response surface methodology, *Int. J. Food Sci. Technol.*, 26, 403, 1991.

Bauman, H., HACCP: concept, development, and application, *Food Technol.*, 44, 156, May 1990.

Baxter, J., Blood, R. M., and Gibbs, P. A., Assessment of antimicrobial effects of lactic acid bacteria, *Br. Food Manuf. Ind. Res. Assoc. Res. Rep.*, No. 425, 1983.

Beauchamp, G. K., Research in chemosensation related to flavor and fragrance perception, *Food Technol.*, 44, 98, Jan. 1990.

Beckers, H. J., Incidence of foodborne diseases in The Netherlands: annual summary 1982 and an overview from 1979 to 1982, *J. Food Prot.*, 51, 327, 1988.

Bender, F. E., Douglas, L. W., and Kramer, A., *Statistical Methods for Food and Agriculture*, AVI Publishing Co., Westport, CT, 1982, chap. 3.

Bengtsson, N., Contact grilling, in *Proceedings of IUFoST International Symposium on Progress in Food Preparation Processes*, SIK — Swedish Food Institute, Göteborg, 1986, 129.

Berger, K. G., Tropical oils in the U.S.A.: situation report, *Food Sci. Technol. Today*, 3, 232, 1989.

Berner, L. A., McBean, L. D., and Lofgren, P. A., Calcium and chronic disease prevention: challenges to the food industry, *Food Technol.*, 44, 50, Mar. 1990.

Bertin, O., Labatt stumbled on to low-cal "fat", *Globe and Mail (Toronto)*, B18, Aug. 3, 1991.

Best, D., Technology ripens opportunities for fruit and vegetable processors, *Prep. Foods*, 157, 83, No. 11, 1988.

Best, D., Marketing technology through the looking glass, *Act. Rep. Res. Dev. Assoc.*, 41, 86, No. 1, 1989.

Best, D., Analogues restructure their market, *Prep. Foods*, 158, 72, No. 11, 1989.

Best, D., New perspectives on water's role in formulation, *Prep. Foods*, 161, 59, No. 9, 1992.

Best, D. and O'Donnell, C. D., Food products for the next millenium, *Prep. Foods*, 161, 48, No. 7, 1992.

Beuchat, L. R., Sensitivity of *Vibrio parahaemolyticus* to spices and organic acid, *J. Food Sci.*, 41, 899, 1976.

Beuchat, L. and Golden, D., Antimicrobials occurring naturally in foods, *Food Technol.*, 43, 134, Jan. 1989.

Binkerd, E. F., The luxury of new product development, *Food Can.*, 35, 31, Nov. 1975.

Bishop, D. G., Spratt, W. A., and Paton, D., Computer plotting in 3-dimensions: a program designed for food science applications, *J. Food Sci.*, 46, 1938, 1981.

Biss, C. H., Coombes, S. A., and Skudder, P. J., The development and application of ohmic heating for the continuous heating of particulate foodstuffs, in *Process Engineering in the Food Industry*, Field, R. W. and Howell, J. A., Eds., Elsevier, London, 1989, 17.

Blanchfield, J. R., How the new food product is designed, *Food Sci. Technol. Today*, 2, 54, 1988.

Bogaty, H., Development of new consumer products — ways to improve your chances, *Res. Manage.*, 26, Jul. 1974.

Bolaffi, A. and Lulay, D., The foodservice industry: continuing into the future with an old friend, *Food Technol.*, 43, 258, Sept. 1989.

Bond, S., New products in the market place, *Food Technol. Int. Eur.*, 109, 1992.

Brackett, R. E., Microbiological safety of chilled foods: current issues, *Trends Food Sci. Technol.*, 3, 81, 1992.

Bradbury, F. R., McCarthy, M. C., and Suckling, C. W., Patterns of innovation: part I, *Chem. Ind.*, 22, 1972.

Bristol, P., Packaging freshness, *Food Can.*, 50, 30, May 1990.

Brooker, J. R. and Nordstrom, R. D., Developments in engineered seafoods for commercial and military markets, *Act. Rep. Res. Dev. Assoc.*, 39, 56, No. 2, 1987.

Brücher, H., Amaranth, an old Amerindian crop, *DRAGOCO Rep.*, 35, No. 2, 1983.

Bruhn, C. and Schutz, H. G., Consumer awareness and outlook for acceptance of food irradiation, *Food Technol.*, 43, 93, Jul. 1989.

Buchanan, R. L., Predictive food microbiology, *Trends Food Sci. Technol.*, 4, 6, 1993.

Burch, N. L. and Sawyer, C., Hospital foodservice requirements: special diet convenience foods, *Food Technol.*, 40, 131, Jul. 1986.

Busch, L., Biotechnology: consumer concerns about risks and values, *Food Technol.*, 45, 96, Apr. 1991.

Bush, P., Expert systems: a coalition of minds, *Prep. Foods*, 158, 162, No. 9, 1989.

Buxton, A., Going online, *Trends Food Sci. Technol.*, 2, 266, 1991.

Campbell, I., Consumer attitudes to food safety, *Visions*, 2, No. 4, 1991.

Campbell, L. B. and Pavlasek, S. J., Dairy products as ingredients in chocolate and confections, *Food Technol.*, 41, 78, Oct. 1987.

Cardoso, G. and Labuza, T. P., Prediction of moisture gain and loss for packaged pasta subjected to a sine wave temperature/humidity environment, *J. Food Technol.*, 18, 587, 1983.

Carlin, F., Nguyen-The, C., Chambroy, Y., and Reich, M., Effects of controlled atmospheres on microbial spoilage, electrolyte leakage and sugar content of fresh "ready-to-use" grated carrots, *Int. J. Food Sci. Technol.*, 25, 110, 1991.

Chapman, S. and McKernan, B. J., Heat conduction into plastic food containers, *Food Technol.*, 17, 79, Sept. 1963.

Chinachoti, P., Water mobility and its relation to functionality of sucrose-containing food systems, *Food Technol.*, 47, 134, Jan. 1993.

Chirife, J. and Favetto, G. J., Some physico-chemical basis of food preservation by combined methods, *Food Res. Int.*, 25, 389, 1992.

Chuzel, G. and Zakhia, N., Adsorption isotherms of gari for estimation of packaged shelf-life, *Int. J. Food Sci. Technol.*, 26, 583, 1991.

Clarke, D., Chilled foods — the caterer's viewpoint, *Food Sci. Technol. Today,* 4, 227, 1990.

Clausi, A. S., The story of dry cereal and freeze dried fruit, presented at School of Food Science, Cornell University, Ithaca, NY, April 21, 1971.

Clausi, A. S., The role of technical research in product development programs, presented at Institute of Food Technologists Short Course on Ingredient Technology for Product Development, New Orleans, LA, May 12–15, 1974.

Cocup, R. O. and Sanderson, W. B., Functionality of dairy ingredients in bakery products, *Food Technol.,* 41, 86, Oct. 1987.

Cohen, J. C., Applications of qualitative research for sensory analysis and product development, *Food Technol.,* 44, 164, Nov. 1990.

Cohen, L. A., Diet and cancer, *Sci. Am.,* 257, 42, Nov. 1987.

Colby, M. and Savagian, J., Irradiation: progress or peril? Con: Consumers say no!, *Prep. Foods,* 158, 62, No. 9, 1989.

Cole, M. B., Databases in modern food microbiology, *Trends Food Sci. Technol.,* 2, 293, 1991.

Cooper, L., A computer system for consumer complaints, *Food Technol. Int. Eur.,* 247, 1990.

Coppock, J. B. M., Has food technology outstripped food science?, *Inst. Food Sci. Technol. Proc.,* 11, 193, 1978.

Cremers, H. C., Fishy cure for craggy features, *New Sci.,* 137, 8, No. 1863, 1993.

Curiale, M. S., Shelf-life evaluation analysis, *Dairy, Food Environ. Sanitation,* 11, 364, 1991.

Daeschel, M. A., Antimicrobial substances from lactic acid bacteria for use as food preservatives, *Food Technol.,* 43, 164, Jan. 1989.

Daniel, S. R., How to develop a customer complaint feedback system, *Food Technol.,* 38, 41, Sept. 1984.

Daniels, R., President, Audits International, personal communication, 1993.

Davies, A., Kochar, S. P., and Weir, G. S., Studies on the efficacy of natural antioxidants, *Leatherhead Food Res. Assoc. Tech. Circ.,* No. 695, 1979.

Davies, R., Birch, G. G., and Parker, K. J., Eds., *Intermediate Moisture Foods,* Applied Science Publishers, Essex, U.K., 1976.

Day, B., Extension of shelf-life of chilled foods, *Eur. Food Drink Rev.,* 47, Autumn 1989.

Day, B., A perspective of modified atmosphere packaging of fresh produce in Western Europe, *Food Sci. Technol. Today,* 4, 215, 1990.

Dean, R. C., The temporal mismatch — innovation's pace vs management's time horizon, *Res. Manage.,* 12, May 1974.

Decareau, R. V., Cooking by microwaves, in *Proc. IUFoST Int. Symp. Progress in Food Preparation Processes,* SIK — Swedish Food Institute, Göteborg, 1986, 173.

Dempster, J. F., Hawrysh, Z. J., Shand, P., Lahola-Chomiak, L., and Corletto, L., Effect of low-dose irradiation (radurization) on the shelf life of beefburgers stored at 3°C, *J. Food Technol.,* 20, 145, 1985.

Dethmers, A. E., Utilizing sensory evaluation to determine product shelf life, *Food Technol.,* 33, 40, Sept. 1979.

Downer, L., *Japanese Vegetarian Cooking,* Pantheon Books, New York, 1986.

Duthie, G. G., Antioxidant hypothesis of cardiovascular disease, *Trends Food Sci. Technol.,* 2, 205, 1991.

Duxbury, D., New products, flavors, and uses of under-utilized species, *Act. Rep. Res. Dev. Assoc.*, 39, 51, No. 2, 1987.

Dziezak, J. D., Fats, oils and fat substitutes, *Food Technol.*, 43, 66, Jul. 1989.

Dziezak, J. D., Taking the gamble out of product development, *Food Technol.*, 44, 110, Jun. 1990.

Dziezak, J. D., A focus on gums, *Food Technol.*, 45, 116, Mar. 1991.

Eisner, M., *Introduction into the Technique and Technology of Rotary Sterilization*, 2nd ed., private author's edition, copyright 1988, M. Eisner, Hafenducht 3, D-2211 Brockdorf, Germany.

Engel, C., Natural colours. Their stability and application in food, *Br. Food Manuf. Ind. Res. Assoc. Sci. Tech. Surv.*, No. 117, 1979.

Erickson, D., Brain, food, *Sci. Am.*, 265, 124, Nov. 1991.

Ewaidah, E. H. and Hassan, B. H., Prickly pear sheets: a new fruit product, *Int. J. Food Sci. Technol.*, 27, 353, 1992.

Farber, J. M., Foodborne pathogenic microorganisms: characteristics of the organisms and their associated diseases. I. Bacteria, *Can. Inst. Food Sci. Technol. J.*, 22, AT/311, 1989.

Farr, D., High pressure technology in the food industry, *Trends Food Sci. Technol.*, 1, 14, 1990.

Fernandez de Tonella, M. L., Taylor, R. R., and Stull, J. W., Properties of a chocolate-flavored beverage from chick-pea, *Cereal Foods World*, 26, 528, 1981.

Figueiredo, A. A., Mesquite: history, composition and food uses, *Food Technol.*, 44, 118, Nov. 1990.

Floros, J. D. and Chinnan, M. S., Computer graphics-assisted optimization for product and process development, *Food Technol.*, 42, 72, Feb. 1988.

Food Development Division, Modified atmosphere packaging: (A) An extended shelf life technology; (B) Investment decisions; (C) The consumer perspective, Agricultural Development Branch, Agriculture Canada, 930 Carling Ave, Ottawa K1A 0C5, 1990.

Francis, F. J., Natural food colorants, *Cereal Foods World*, 26, 565, 1981.

Francis, F. J., A new group of food colorants, *Trends Food Sci. Technol.*, 3, 27, 1992.

Friedman, M., Twenty-five years and 98,900 new products later ..., *Prep. Foods*, 159, 23, No. 8, 1990.

Gabriel, S. L., Separation and Identification of the Anthocyanins in the Sweet Potato (*Ipomoea batatas*) using High Pressure Liquid Chromatographic Methods, Master's thesis, Food Science and Nutrition, University of Massachusetts, 1989.

Gains, N. and Thomson, D., Contextual evaluation of canned lagers using repertory grid method, *Int. J. Food Sci. Technol.*, 25, 699, 1990.

Gaunt, I. F., Food irradiations — safety aspects, *Inst. Food Sci. Technol. Proc.*, 19, 171, 1985.

Geake, E. and Coghlan, A., Industry "does not need research", *New Sci.*, 133, 15, No. 1808, 1992.

Geeson, J. D., Smith, S. M., Everson, H. P., George, P. M., and Browne, K. M., Modified atmosphere packaging to extend the shelf life of tomatoes, *Int. J. Food Sci. Technol.*, 22, 659, 1987.

Geeson, J. D., Genge, P., Smith, S. M., and Sharples, R. O., The response of unripe Conference pears to modified atmosphere retail packaging, *Int. J. Food Sci. Technol.*, 26, 215, 1991.

Gélinas, P., Freezing and fresh bread, *Alimentech*, 4, 12, June 1991.

Gibbons, M., Greer, J. R., Jevons, F. R., Langrish, J., and Watkins, D. S., Value of curiosity-oriented research, *Nature*, 225, 1005, 1970.

Gibbs, P. A. and Williams, A. P., Using mathematics for shelf life prediction, *Food Technol. Int. Eur.*, 287, 1990.

Giddings, G. G., Sterilization of spices: irradiation vs gaseous sterilization, *Act. Rep. Res. Dev. Assoc.*, 36, 20, No. 2, 1984.

Giddings, G. G., Irradiation: progress or peril? Pro: Safety is no longer an issue, *Prep. Foods*, 158, 62, No. 9, 1989.

Giese, J. H., Alternative sweeteners and bulking agents, *Food Technol.*, 47, 114, Jan. 1993.

Gitelman, P., Opportunities in the food industry, *Can. Inst. Food Sci. Technol. J.*, 19, XIX, 1986.

Glew, G., Introduction to catering production planning, in *Proc. IUFoST Int. Symp. Progress in Food Preparation Processes*, SIK — Swedish Food Institute, Göteborg, 1986, 203.

Godfrey, W., A retailing perspective, *Food Sci. Technol. Today, 2*, 56, 1988.

Goff, H. D., Low-temperature stability and the glassy state in frozen foods, *Food Res. Int.*, 25, 317, 1992.

Goldenfield, I., The regulations affecting product development — industry view, *Food Technol.*, 31, 80, Jul. 1977.

Goldman, A., A study of product development management practice among food manufacturing companies located in southern Ontario, Working Paper No. 84-303, ISSN 0826-8878, Dept. of Consumer Studies, University of Guelph, Ontario, 1983.

Goldman, A., personal communication, 1993.

Gordon, M. H., Finding a role for natural antioxidants, *Food Technol. Int. Eur.*, 187, 1989.

Gould, G., Predictive mathematical modelling of microbial growth and survival in foods, *Food Sci. Technol. Today*, 3, 89, 1989.

Graf, E. and Saguy, I. S., *Food Product Development: From Concept to the Marketplace*, Graf, E. and Saguy, I. S., Eds., Van Nostrand Reinhold, New York, 1991, chap. 3.

Graham, D., Quality programs and consumer complaints, *Food Technol. Int. Eur.*, 245, 1990.

Grodner, R. and Hinton, A., Jr., Low dose gamma irradiation of *Vibrio cholerae* in crabmeat (*Callinectes sapidus*), in *Proc. of Eleventh Annual Tropical and Subtropical Fisheries Conf. of the Americas*, Grodner, R. and Hinton, A., Jr., Eds., Texas A&M University, College Station, 1986, 219.

Gutteridge, C. S., New methods for finding the right market niche, *Food Technol. Int. Eur.*, 127, 1990.

Halden, K., De Alwis, A. A. P., and Fryer, P. J., Changes in the electrical conductivity of foods during ohmic heating, *Int. J. Food Sci. Technol.*, 25, 9, 1990.

Hamm, D. J., Preparation and evaluation of trialkoxytricarbalkylate, trioxycitrate, trialkoxyglycerylether, jojoba oil and sucrose polyester as low calorie replacements of edible fats and oils, *J. Food Sci.*, 49, 419, 1984.

Hannigan, K. J., Dried citrus juice sacs add moisture to food products, *Food Eng.*, 54, 88, Mar. 1982.

Hardy, K. G., Fickle tastebuds, *Bus. Q.*, 40, Spring 1991.

Harlander, S., Food technology: yesterday, today, and tomorrow, *Food Technol.*, 43, 196, Sept. 1989.

Harlander, S., Social, moral and ethical issues in food biotechnology, *Food Sci. Technol. Today*, 6, 66, 1992.

Hauck, K., Alone on the range, *Prep. Foods*, 161, 32, No. 11, 1992.

Hayashi, R., Application of high pressure to food processing and preservation: philosophy and development, in *Engineering and Food*, Vol. 2, Spiess, W. E. L. and Schubert, H., Eds., Elsevier Applied Science, London, 1989, 815.

Head, A. W., The technical manager: present and future role, *Chem. Ind.*, 716, 1971.

Heath, H., *Flavor Technology: Profiles, Products, Applications*, AVI Publishing Company, Inc., Westport, CT, 1978.

Henika, R. G., Simple and effective system for use with response surface methodology, *Cereal Sci. Today*, 17, 309, 1972.

Herrod, R. A., Industrial applications of expert systems and the role of the knowledge engineer, *Food Technol.*, 43, 130, May 1989.

Hibler, M., Fast foods, *Can. Consumer*, 18, 19, Jul. 1988.

Hill, C. G., Jr. and Grieger-Block, R. A., Kinetic data: generation, interpretation, and use, *Food Technol.*, 34, 56, Feb. 1980.

Hill, S., The IFIS food science and technology bibliographic databases, *Trends Food Sci. Technol.*, 2, 269, 1991.

Holmes, A. W., The control of research for profit, *Br. Food Manuf. Ind. Res. Assoc. Tech. Circ.*, No. 412, 1968.

Holmes, A. W., Securing innovation in the food industry, *Br. Food Manuf. Ind. Res. Assoc. Tech. Circ.*, No. 636, 1977.

Hoover, D. G., Metrick, C., Papineau, A. M., Farkas, D., and Knorr, D., Biological effects of high hydrostatic pressure on food microorganisms, *Food Technol.*, 43, 99, Mar. 1989.

Hsieh, Y. P. C., Pearson, A. M., and Magee, W. T., Development of a synthetic meat flavor mixture by using surface response methodology, *J. Food Sci.*, 45, 1125, 1980.

Hughes, D. B. and Hoover, D. G., Bifidobacteria: their potential for use in American dairy products, *Food Technol.*, 45, 74, Apr. 1991.

Huizenga, T. P., Liepins, K., and Pisano, D. J., Jr., Early involvement, *Qual. Prog.*, 81, Jun. 1987.

Idziak, E. S., personnal communication, 1993.

James, W., *The Principles of Psychology*, 1890, chap. 2. Quoted in *Bartlett's Familiar Quotations*, 15th and 25th edition, p. 649. Little, Brown and Company, Toronto, Canada, 1980.

Jenkins, D., Dietary fibre and its relation to nutrition and health, *Inst. Food Sci. Technol. Proc.*, 13, 51, 1980.

Jeremiah, L. E., Penney, N., and Gill, C. O., The effects of prolongued storage under vacuum or CO_2 on the flavor and texture profiles of chilled pork, *Food Res. Int.*, 25, 9, 1992.

Jezek, E. and Smyrl, T. G., Volatile changes accompanying dehydration of apples by the Osmovac process, *Can. Inst. Food Sci. Technol. J.*, 13, 43, 1980.

Johnston, W. A., Surimi — an introduction, *Eur. Food Drink Rev.*, 21, Autumn 1989.

Jolly, D. A., Schutz, H. G., Diaz-Knauf, K. V., and Johal, J., Organic foods: consumer attitudes and use, *Food Technol.*, 43, 60, Nov. 1989.

Jones, K., The EEC flavoring and food labelling directives, *DRAGOCO Rep.*, 103, No. 3, 1992.

Josephson, E. S., Military benefits of food irradiation, *Act. Rep. Res. Dev. Assoc.*, 36, 30, No. 2, 1984.

Juven, B. J., Schved, F., and Linder, P., Antagonistic compounds produced by a chicken intestinal strain of *Lactobacillus acidophilus*, *J. Food Prot.*, 55, 157, 1992.

Kantor, D., New product proliferation: are the benefits worth the cost?, *Prep. Foods*, 160, 28, No. 7, 1991.

Katzenstein, A. W., The Food Update Delphi Survey: forecasting the food industry 10 years from now, *Food Prod. Dev.*, 11, Jun. 1975.

Kennedy, J. P., Structured lipids: fats of the future, *Food Technol.*, 45, 76, Nov. 1991.

Kernon, J. M., The Foodline scientific and technical, marketing and legislation databases, *Trends Food Sci. Technol.*, 2, 276, 1991.

King, V. A.-E. and Zall, R. R., A response surface methodology approach to the optimization of controlled low-temperature vacuum dehydration, *Food Res. Int.*, 25, 1, 1992.

Kirk, D. and Osner, R. C., Collection of data on food waste from catering outlets in a University and a Polytechnic, *Inst. Food Sci. Technol. Proc.*, 14, 190, 1981.

Kirkpatrick, K. J. and Fenwick, R. M., Manufacture and general properties of dairy ingredients, *Food Technol.*, 41, 58, Oct. 1987.

Kläui, H., Naturally occurring antioxidants, *Inst. Food Sci. Technol. Proc.*, 6, 195, 1973.

Klensin, J. C., Information technology and food composition databases, *Trends Food Sci. Technol.*, 2, 279, 1991.

Knorr, D., Recovery and utilization of chitin and chitosan in food processing waste management, *Food Technol.*, 45, 114, Jan. 1991.

Knorr, D., Beaumont, M. D., Caster, C. S., Dörnenburg, H., Gross, B., Pandya, Y., and Romagnoli, L. G., Plant tissue culture for the production of naturally derived food ingredients, *Food Technol.*, 44, 71, Jun. 1990.

Kolata, G., Food affects human behaviour, *Science*, 218, 1209, 1982.

Konuma, H., Shinagawa, K., Tokumaru, M., Onove, Y., Konno, S., Fujino, N., Shigehisa, T., Kurata, H., Kuwabara, Y., and Lopes, C. A. M., Occurrence of *Bacillus cereus* in meat products, raw meat and meat product additives, *J. Food Prot.*, 51, 324, 1988.

Kritchevsky, D., The effect of dietary garlic on the development of cardiovascular disease, *Trends Food Sci. Technol.*, 2, 141, 1991.

Krizmanic, J., Here's who we are!, *Veg. Times*, 182, 72, 1992.

Kroll, B. J., Evaluating rating scales for sensory testing with children, *Food Technol.*, 44, 78, Nov. 1990.

LaBarge, R. G., The search for a low-caloric oil, *Food Technol.*, 42, 84, Jan. 1988.

Labuza, T. P., The effect of water activity on reaction kinetics of food deterioration, *Food Technol.*, 34, 36, Apr. 1980.

Labuza, T. P. and Riboh, D., Theory and application of Arrhenius kinetics to the prediction of nutrient losses in foods, *Food Technol.*, 36, 66, Oct. 1982.

Labuza, T. P. and Schmidl, M. K., Accelerated shelf-life testing of foods, *Food Technol.*, 39, 57, Sept. 1985.

Land, E., The second great product of industry: the rewarding working life, presented at Science and Human Progress: 50th Anniversary of Mellon Institute, Pittsburgh, PA, 1963, 107.

Lanier, T. C., Functional properties of surimi, *Food Technol.*, 40, 107, Mar. 1986.

Lechowich, R. V., Microbiological challenges of refrigerated foods, *Food Technol.*, 42, 84, Dec. 1988.

Lee, C. M., Surimi process technology, *Food Technol.*, 38, 69, Nov. 1984.

Lee, J., The development of new products for the European Market, *Food Sci. Technol. Today*, 5, 155, 1991.

Lee, K., Food neophobia: major causes and treatment, *Food Technol.*, 43, 62, Dec. 1989.

Leistner, L., Hurdle technology applied to meat products of the shelf stable product and intermediate moisture food types, in *Properties of Water in Foods*, Simatos, D. and Multon, J. L., Eds., Martinus Nijhoff Publishers, Dordrecht, 1985, 309.

Leistner, L., Shelf stable products and intermediate moisture foods based on meat, presented at IFT-IUFoST Basic Symp. Water Activity: Theory and Applications, Dallas, June 13 to 14, 1986, chap. 13.

Leistner, L., Food preservation by combined methods, *Food Res. Int.*, 25, 151, 1992.

Leistner, L. and Rödel, W., Inhibition of micro-organisms in food by water activity, in *Inhibition and Inactivation of Vegetative Microbes*, Skinner, F. A. and Hugo, W. B., Eds., Academic Press, London, 1976a, 219.

Leistner, L. and Rödel, W., The stability of intermediate moisture foods with respect to micro-organisms, in *Intermediate Moisture Foods*, Davies, R., Birch, G. G., and Parker, K. J., Eds., Applied Science Publishers, London, 1976b, chap. 10.

Leistner, L., Rödel, W., and Krispien, K., Microbiology of meat and meat products in high- and intermediate-moisture ranges, in *Water Activity: Influences on Food Quality*, Rockland, L. B. and Stewart, G. F., Eds., Academic Press, New York, 1981, 855.

Lenz, M. K. and Lund, D. B., Experimental procedures for determining destruction kinetics of food components, *Food Technol.*, 34, 51, Feb. 1980.

Lieber, H., U.S. Patent 788.480, Preserved radio-active organic matter and food, 1905.

Light, N., Young, H., and Youngs, A., Operating temperatures in chilled food vending machines and risk of growth of food poisoning organisms, *Food Sci. Technol. Today*, 1, 252, 1987.

Lightbody, M. S., New technological approaches to reducing uniformity in processed foods, *Food Sci. Technol. Today Proc.*, 4, 37, 1990.

Lingle, R., AmeriQual Foods: at ease with MRE's, *Prep. Foods*, 158, 144, No. 9, 1989.

Lingle, R., Degradable plastics: all sizzle and no steak?, *Prep. Foods*, 159, 144, Jan. 1990.

Lingle, R., Streamlining package design through computers, *Prep. Foods*, 160, 86, No. 8, 1991.

Lingle, R., A sign of changing times, *Prep. Foods*, 161, 52, No. 2, 1992.

Livingston, G. E., Foodservice: older than Methuselah, *Food Technol.*, 44, 54, Jul. 1990.

Loaharanu, P., International trade in irradiated foods: regional status and outlook, *Food Technol.*, 43, 77, Jul. 1989.

Lund, D. B., Considerations in modeling food processes, *Food Technol.*, 37, 92, Jan. 1983.

MacFie, H., Factors affecting consumers' choice of food, *Food Technol. Int. Eur.*, 123, 1990.

Malpas, R., Chemical technology — scaling greater heights in the next ten years?, *Chem. Ind.*, 111, 1977.

Mans, J., Kyotaru's bridge across the Pacific, *Prep. Foods*, 161, 85, No. 12, 1992.

Marcotte, M., Irradiated strawberries enter the U.S. market, *Food Technol.*, 46, 80, May 1992.

Mardon, J., Cripps, W. C., and Matthews, G. T., The organisation and administration of technical departments in large multi-plant companies, *Chem. Ind.*, 450, 1970.

Marlow, P., Qualitative research as a tool for product development, *Food Technol.*, 41, 74, Nov. 1987.

Martin, D., The impact of branding and marketing on perception of sensory qualities, *Food Sci. Technol. Today Proc.*, 4, 44, 1990.

Mason, L. H., Church, I. J., Ledward, D. A., and Parsons, A. L., The sensory quality of foods produced by conventional and enhanced cook-chill method, *Int. J. Food Sci. Technol.*, 25, 247, 1990.

Matthews, M. E., Foodservice in health care facilities, *Food Technol.*, 36, 53, Jul. 1982.

Mattson, P., Eleven steps to low cost product development, *Food Prod. Dev.*, 6, 106, 1970.

Maugh, T. H., II, "Cancer is not inevitable", *Science*, 212, 36, 1982.

Mayo, J. S., AT&T: management questions for leadership in quality, *Qual. Prog.*, 34, Apr. 1986.

Mazza, G., Development and consumer evaluation of a native fruit product, *Can. Inst. Food Sci. Technol. J.*, 12, 166, 1979.

McDermott, B. J., Identifying consumers and consumer test subjects, *Food Technol.*, 44, 154, Nov. 1990.

McDermott, R. L., Functionality of dairy ingredients in infant formula and nutritional specialty products, *Food Technol.*, 41, 91, Oct. 1987.

McGinn, C. J. P., Evaluation of shelf life, *Inst. Food Sci. Technol. Proc.*, 15, 153, 1982.

McLellan, M. R., An introduction to artificial intelligence and expert systems, *Food Technol.*, 43, 120, May 1989.

McLellan, M. R., Hoo, A. F., and Peck, V., A low-cost computerized card system for the collection of sensory data, *Food Technol.*, 41, 66, Nov. 1987.

McPhee, M., Organically grown-up?, *Prep. Foods*, 161, 17, No. 3, 1992.

McProud, L. M. and Lund, D. B., Thermal properties of beef loaf produced in foodservice systems, *J. Food Sci.*, 48, 677, 1983.

McWatters, K. H., Resurreccion, A. V. A., and Fletcher, S. M., Response of American consumers to akara, a traditonal West African food made from cowpea paste, *Int. J. Food Sci. Technol.*, 25, 551, 1990.

McWatters, K. H., Enwere, N. J., and Fletcher, S. M., Consumer response to akara (fried cowpea paste) served plain or with various sauces, *Food Technol.*, 46, 111, Feb. 1992.

McWeeny, D. J., Long term storage of some dry foods: a discussion of some of the principles involved, *J. Food Technol.*, 15, 195, 1980.

Megremis, C. J., Medium-chain triglycerides: a nonconventional fat, *Food Technol.*, 45, 108, Feb. 1991.

Meisel, H. and Schlimme, E., Milk proteins: precursors of bioactive peptides, *Trends Food Sci. Technol.*, 1, 41, Aug. 1990.

Meltzer, R., Value added products: a noteworthy niche, *Visions*, 2, No. 3, 1991.

Mermelstein, N. H., Retort pouch earns 1978 IFT Food Technology Industrial Achievement Award, *Food Technol.*, 32, 22, Jun. 1978.

Mertens, B. and Knorr, D., Developments of nonthermal processes for food preservation, *Food Technol.*, 46, 124, May 1992.

Metrick, C., Hoover, D. G., and Farkas, D. F., Effects of high hydrostatic pressure on heat-resistant and heat-sensitive strains of *Salmonella*, *J. Food Sci.*, 54, 1547, 1989.

Meyer, R. S., Eleven stages of successful new product development, *Food Technol.*, 38, 71, Jul. 1984.

Mills, E. N. C., Alcocer, M. J. C., and Morgan, M. R. A., Biochemical interactions of food-derived peptides, *Trends Food Sci. Technol.*, 3, 64, 1992.

Moskowitz, H. R., Benzaquen, I., and Ritacco, G., What do consumers really think about your product?, *Food Eng.*, 53, 80, Sept. 1981.

Mossel, D. A. A. and Ingram, M., The physiology of the microbial spoilage of food, *J. Appl. Bacteriol.*, 18, 232, 1955.

Mullen, K. and Ennis, D. M., Rotatable designs in product development, *Food Technol.*, 33, 74, Jul. 1979.

Mullen, K. and Ennis, D. M., Mathematical system enters development realm as aid in achieving optimum ingredient levels, *Food Prod. Dev.*, 15, 50, Nov. 1979.

Mullen, K. and Ennis, D., Fractional factorials in product development, *Food Technol.*, 39, 90, May 1985.

Muller, R. A., Innovation and scientific funding, *Science*, 209, 880, 1980.

Mundy, C. C., Accessing the literature of food science, *Trends Food Sci. Technol.*, 2, 272, 1991.

Nazario, S. L., Big firms get high on organic farming, *Wall Street Journal (New York)*, B1, Mar. 21, 1989.

Neaves, P., Gibbs, P. A., and Patel, M., Inhibition of *Clostridium botulinum*, by preservative interactions, *Br. Food Manuf. Ind. Res. Assoc. Res. Rep.*, No. 378, 1982.

Newsome, R., Organically grown foods, *Food Technol.*, 44, 26, Jun. 1990.

Noel, T. R., Ring, S. G., and Whittam, M. A., Glass transitions in low-moisture foods, *Trends Food Sci. Technol.*, 1, 62, 1990.

Norback, J., Techniques for optimization of food processes, *Food Technol.*, 34, 86, Feb. 1980.

Normand, F. L., Ory, R. L., and Mod, R. R., Binding of bile acids and trace minerals by soluble hemicelluloses of rice, *Food Technol.*, 41, 86, Feb. 1987.

Norwig, J. F. and Thompson, D. R., Making accurate energy efficiency measurements in foodservice equipment, *Act. Rep. Res. Dev. Assoc.*, 36, 37, No. 2, 1984.

O'Brien, J., An overview of online information resources for food research, *Trends Food Sci. Technol.*, 2, 301, 1991.

O'Donnell, C., Computers: what's the use?, *Prep. Foods*, 160, 62, No. 8, 1991.

O'Donnell, C. D., Aphrodisiacs, *Prep. Foods*, 162, 69, No. 3, 1992.

Ohlsson, T., Boiling in water, in *Proc. IUFoST Int. Symp. Progress in Food Preparation Processes*, SIK — Swedish Food Institute, Göteborg, 1986, 89.

Oickle, J. G., *New Product Development and Value Added*, Food Development Division, Agriculture Canada, May 1990.

Okamoto, M., Kawamura, Y., and Hayashi, R., Application of high pressure to food processing: textural comparison of pressure- and heat-induced gels of food proteins, *Agric. Biol. Chem.*, 54, 183, No. 1, 1990.

Okoli, E. C. and Ezenweke, L. O., Formulation and shelf-life of a bottled pawpaw juice beverage, *Int. J. Food Sci. Technol.*, 25, 706, 1990.

Olson, A., Gray, G. M., and Chiu, M., Chemistry and analysis of soluble dietary fiber, *Food Technol.*, 41, 71, Feb. 1987.

O'Neill, M., Geneticists' latest discovery: public fear of "Frankenfood", *N. Y. Times*, 1, Jun. 28, 1992.

O'Sullivan, M. G., Thornton, G., O'Sullivan G. C., and Collins, J. K., Probiotic bacteria: myth or reality?, *Trends Food Sci. Technol.*, 3, 309, 1992.

Park, C. E., Szabo, R., and Jean, A., A survey of wet pasta packaged under a CO:N (20:80) mixture for Staphylococci and their enteriotoxins, *Can. Inst. Food Sci. Technol. J.*, 21, 109, 1988.

Parsons, R., *Statistical Analysis: A Decision-Making Approach*, 2nd ed., Harper & Row, New York, 1978, chap. 5 and chap. 6.

Paster, N., Juven, B. J., Gagel, S., Saguy, I., and Padova, R., Preservation of a perishable pomegranate product by radiation pasteurization, *J. Food Technol.*, 20, 367, 1985.

Patel, T., German syringes turn up in French quarry, *New Sci.*, 135, 7, No. 1835, 1992.

Patel, T., Energy booster "a con" says French food watchdog, *New Sci.*, 137, 8, No. 1863, 1993.

Pearce, S. J., Quality assurance involvement in new product design, *Food Technol.*, 41, 104, Apr. 1987.

Pennington, J. A. T. and Butrum, R. R., Food descriptions using taxonomy and the "Langual" system, *Trends Food Sci. Technol.*, 2, 285, 1991.

Penny, C., Detailing dietary fibre, *Food Ingred. Proc. Int.*, 14, Nov. 1992.

Peryam, D. R., Sensory evaluation — the early days, *Food Technol.*, 44, 86, Jan. 1990.

Peters, J. W., Foodservice R & D at Gino's — spotting the winners faster, *Food Prod. Dev.*, 16, 28, Oct. 1980.

Pine, R. and Ball, S., Productivity and technology in catering operations, *Food Sci. Technol. Today*, 1, 174, 1987.

Pokorný, J., Natural antioxidants for food use, *Trends Food Sci. Technol.*, 2, 223, 1991.

Porter, W. L., Storage life prediction under noncontrolled environmental temperatures: product-sensitive environmental call-out, in *Proc. Food Processors' Institute: Shelf-life: A Key to Sharpening your Competitive Edge*, Food Processors Institute, Washington, D.C., 1981.

Poste, L. M., Mackie, D. B., Butler, G., and Larmond, E., Laboratory Methods for Sensory Analysis of Food, Publication 1864/E, Research Branch, Agriculture Canada, 1991.

Powers, J. J., Uses of multivariate methods in screening and training sensory panelists, *Food Technol.*, 42, 123, Nov. 1988.

Pszczola, D. E., Food irradiation: countering the tactics and claims of opponents, *Food Technol.*, 44, 92, Jun. 1990.

Puzo, D. P., First irradiated fruit on market sells quickly, *Los Angeles Times*, 1986.

Reiser, S., Metabolic effects of dietary pectins related to human health, *Food Technol.*, 41, 91, Feb. 1987.

Righelato, R. C., Biotechnology and food manufacturing, *Food Technol. Int. Eur.*, 155, 1987.

Ripsin, C. M. and Keenan, J. M., The effects of dietary oat products on blood cholesterol, *Trends Food Sci. Technol.*, 3, 137, 1992.

Rizvi, S. S. H. and Acton, J. C., Nutrient enhancement of thermostabilized foods in retort pouches, *Food Technol.*, 36, 105, Apr. 1982.

Roberts, T. A., Combinations of antimicrobials and processing methods, *Food Technol.*, 43, 156, Jan. 1989.

Roberts, T. A., Predictive modelling of microbial growth, *Food Technol. Int. Eur.*, 231, 1990.

Robinson, D. S., Irradiation of foods, *Inst. Food Sci. Technol. Proc., 19*, 165, 1985.

Rockland, L. B. and Nishi, S., Influence of water activity on food product quality and stability, *Food Technol.*, 34, 42, Apr. 1980.

Roos, Y. and Karel, M., Applying state diagrams to food processing and development, *Food Technol.*, 45, 66, Dec. 1991.

Roos, Y. and Karel, M., Amorphous state and delayed ice formation in sucrose solutions, *Int. J. Food Sci. Technol.,* 26, 553, 1991.

Roser, B., Trehalose, a new approach to premium dried foods, *Trends Food Sci. Technol.,* 2, 166, 1991.

Ross, C., personal communication, 1980.

Rowley, D. B., Significance of sublethal injury of foodborne pathogenic and spoilage bacteria during processing, *Act. Rep. Res. Dev. Assoc.*, 36, 41, No. 1, 1984.

Rusoff, I., Nutrition and dietary fiber, *Cereal Foods World*, 29, 668, 1984.

Rutledge, K. R., Accelerated training of sensory descriptive flavor analysis panelists, *Food Technol.*, 46, 114, Nov. 1992.

Rutledge, K. P. and Hudson, J. M., Sensory evaluation: method for establishing and training a descriptive flavor analysis panel, *Food Technol.*, 44, 78, Dec. 1990.

Ryley, J., The impact of fast foods on U.K. nutrition, *Inst. Food Sci. Technol. Proc., 16*, 58, 1983.

Ryval, M., The shape of food to come, *Financ. Post Mag.*, 38, Apr. 15, 1981.

Saca, S. A. and Lozano, J. E., Explosion puffing of bananas, *Int. J. Food Sci. Technol.,* 27, 419, 1992.

Saguy, I. and Karel, M., Models of quality deterioration during food processing and storage, *Food Technol.*, 34, 78, Feb. 1980.

Saito, Y., The retort pouch is a way of life in Japan, *Act. Rep. Res. Dev. Assoc.*, 35, 8, No. 2, 1983.

Saldana, G., Meyer, R., and Lime, B. J., A potential processed carrot product, *J. Food Sci.*, 45, 1444, 1980.

Schmidl, M. K., Massaro, S. S., and Labuza, T. P., Parenteral and enteral food systems, *Food Technol.*, 42, 77, Jul. 1988.

Schneeman, B. O., Soluble vs insoluble fiber — different physiological responses, *Food Technol.*, 41, 81, Feb. 1987.

Schutz, H. G., Multivariate analyses and the measurement of consumer attitudes and perceptions, *Food Technol.*, 42, 141, Nov. 1988.

Scott, V. N., Control and prevention of microbial problems in new generation refrigerated foods, *Act. Rep. Res. Dev. Assoc.*, 39, 22, No. 2, 1987.

Scriven, F. M., Gains, N., Green, S. R., and Thomson, D. M. H., A contextual evaluation of alcoholic beverages using the repertory grid method, *Int. J. Food Sci. Technol.,* 24, 173, 1989.

Selman, J. D., Trends in food processing and food processes, *Food Technol. Int. Eur.,* 55, 1991.

Selman, J., New technologies for the food industry, *Food Sci. Technol. Today,* 6, 205, 1992.

Sensory Evaluation Division, IFT, Sensory Evaluation Guide for Testing Food and Beverage Products, Institute of Food Technologists, Chicago, IL, 1981.

Settlemyer, K., A systematic approach to foodservice new product development, *Food Technol.*, 40, 120, Jul. 1986.

Shapton, D. A. and Shapton, N. F., *Principles and Practices for the Safe Processing of Foods*, Shapton, D. A. and Shapton, N. F., Eds., Butterworth Heinemann Ltd., Oxford, 1991.

Shelef, L. A., Naglik, O. A., and Bogen, D. W., Sensitivity of some common food-borne bacteria to the spices sage, rosemary, and allspice, *J. Food Sci.*, 45, 1042, 1980.

Sherman, H. C., *Food Products*, Macmillan, New York, 1916, 397.

Shi, Z., Bassa, I. A., Gabriel, S. L., and Francis, F. J., Anthocyanin pigments of sweet potatoes — *Ipomoea batatas*, *J. Food Sci.*, 57, 755, 1992a.

Shi, Z., Francis, F. J., and Daun, H., Quantitative comparison of the stability of anthocyanins from *Brassica oleracea* and *Tradescantia pallida* in non-sugar drink model and protein model systems, *J. Food Sci.*, 57, 768, 1992b.

Singhal, R. S., Gupta, A. K., and Kulharni, P. R., Low-calorie fat substitutes, *Trends Food Sci. Technol.*, 2, 241, 1991.

Sinki, G., Technical myopia, *Food Technol.*, 40, 86, Dec. 1986.

Skinner, R. H. and Debling, G. B., Food industry applications of linear programming, *Food Manuf.*, 44, 35, Oct. 1969.

Skjöldebrand, C., Cooking by infrared radiation, in *Proc. IUFoST Int. Symp. Progress in Food Preparation Processes*, SIK — Swedish Food Institute, Göteborg, 1986, 157.

Slade, L. and Levine, H., Beyond water activity: recent advances based on an alternative approach to the assessment of food quality and safety, *Crit. Rev. Food Sci. Nutr.*, 30, 115, 1991.

Slade, P. J., Monitoring *Listeria* in the food production environment. I. Detection of *Listeria* in processing plants and isolation methodology, *Food Res. Int.*, 25, 45, 1992.

Slight, H., The storage and transport of chilled foods — a temperature survey, *Leatherhead Food Res. Assoc. Res. Rep.*, No. 340, 1980.

Snow, P., The shape of shopping to come, *Oxford Today*, 5, 55, No. 1, 1992.

Snyder, O. P., A model food service quality assurance system, *Food Technol.*, 35, 70, Feb. 1981.

Snyder, O. P., Jr., Minimizing foodservice energy consumption through improved recipe engineering and kitchen design, *Act. Rep. Res. Dev. Assoc.*, 36, 49, No. 2, 1984.

Snyder, O. P., Jr., Microbiological quality assurance in foodservice operations, *Food Technol.*, 40, 122, Jul. 1986.

Solberg, M., Buckalew, J. J., Chen, C. M., Schaffner, D. W., O'Neill, K., McDowell, J., Post, L. S., and Boderck, M., Microbiological safety assurance system for foodservice facilities, *Food Technol.*, 44, 68, Dec. 1990.

Stavric, S. and Speirs, J. I., *Escherichia coli* associated with hemorrhagic colitis, *Can. Inst. Food Sci. Technol. J.*, 22, AT/205, 1989.

Stevenson, K. E., Implementing HACCP in the food industry, *Food Technol.*, 44, 179, May 1990.

Stiles, M. E. and Hastings, J. W., Bacteriocin production by lactic acid bacteria: potential for use in meat preservation, *Trends Food Sci. Technol.*, 2, 247, 1991.

Stuller, J., The nature and process of creativity, *Sky*, 11, 37, No. 1, 1982.

Taylor, A. W., Scaling-up of process operations in the food industry, *Inst. Food Sci. Technol. Proc.*, 2, 86, 1969.

Taylor, L. and Leith, S., TQM, *Food Can.*, 51, 18, Jun. 1991.

Taylor, T. G., Diet and coronary heart disease, *Inst. Food Sci. Technol. Proc.*, 13, 45, 1980.

Teutonico, R. A. and Knorr, D., Amaranth: composition, properties, and applications of a rediscovered food crop, *Food Technol.*, 39, 49, Apr. 1985.

Thomson, D., Recent advances in sensory and affective methods, *Food Sci. Technol. Today*, 3, 83, 1989.

Thorne, S., What temperature?, *Inst. Food Sci. Technol Proc.*, 11, 207, 1978.

Tiwari, N. P. and Aldenrath, S. G., Occurrence of *Listeria* species in food and environmental samples in Alberta, *Can. Inst. Food Sci. Technol. J.*, 23, 109, 1990.

Toma, R. B., Curtis, D. J., and Sobotar, C., Sucrose polyester: its metabolic role and possible future applications, *Food Technol.*, 42, 93, Jan. 1988.

Tsuji, K., Low-dose cobalt 60 irradiation for reduction of microbial contamination in raw materials for animal health products, *Food Technol.*, 37, 48, Feb. 1983.

Tung, M. A., Garland, M. R., and Campbell, W. E., Quality comparison of cream style corn processed in rigid and flexible containers, *Can. Inst. Food Sci. Technol. J.*, 8, 211, 1975.

Tuomy, J. M. and Young, R., Retort-pouch packaging of muscle foods for the Armed Forces, *Food Technol.*, 36, 68, Feb. 1982.

Turner, M. and Glew, G., Home-delivered meals for the elderly, *Food Technol.*, 36, 46, Jul. 1982.

Tye, R. J., Konjac flour: properties and applications, *Food Technol.*, 45, 81, Mar. 1991.

van den Hoven, M., Functionality of dairy ingredients in meat products, *Food Technol.*, 41, 72, Oct. 1987.

Waite-Wright, M., Chilled foods — the manufacturer's responsibility, *Food Sci. Technol. Today*, 4, 223, 1990.

Walker, S. J. and Jones, J. E., Predictive microbiology: data and model bases, *Food Technol. Int. Eur.*, 209, 1992.

Walston, J., C.O.D.E.X. spells controversy, *Ceres*, 24, 28, Jul./Aug. 1992.

Webster, S. N., Fowler, D. R., and Cooke, R. D., Control of a range of food related microorganisms by a multi-parameter preservation system, *J. Food Technol.*, 20, 311, 1985.

Whitehead, R., What'll we eat in 1999?, *Ind. Week*, 30, May 17, 1976.

Whitney, L. F., What expert systems can do for the food industry, *Food Technol.*, 43, 135, May 1989.

Wiley, H. W., *Foods and their Adulteration*, 3rd ed., P. Blakiston's Sons & Co., Philadelphia, PA, 1917, 455.

Wilhelmi, F., Product safety and how to ensure it, *DRAGOCO Rep.*, 14, No. 1, 1988.

Williams, A. P., Blackburn, C., and Gibbs, P., Advances in the use of predictive techniques to improve the safety and extend the shelf-life of foods, *Food Sci. Technol. Today*, 6, 148, 1992.

Williams, E. F., Weak links in the cool and cold chain, *Inst. Food Sci. Technol. Proc.*, 11, 211, 1978.

Williams, M., Electronic databases, *Science*, 228, 445, 1985.

Williams, R., Vending in the Canadian foodservice industry, *Visions*, 2, No. 5, 1991.

Williams, R., A profile of the Canadian fast-food sector, *Visions*, 3, No. 2, 1992.

Williamson, M., Tasting tests carried out at the Leatherhead Food Research Association, *Leatherhead Food Res. Assoc. Tech. Circ.*, No. 749, 1981.

Wilson, C., Ethical shopping, *Food Can.*, 52, 7, Nov./Dec. 1992.

Wolfe, K. A., Use of reference standards for sensory evaluation of product quality, *Food Technol.*, 33, 43, Sept. 1979.

Wood, P. J., Oat β-glucan — physicochemical properties and physiological effects, *Trends Food Sci. Technol.*, 2, 311, 1991.

Wood, S., Food law enforcement — where next? An industry viewpoint, *Inst. Food Sci. Technol. Proc.*, 18, 89, 1985.

Wurtman, R. J. and Wurtman, J. J., Carbohydrates and depression, *Sci. Am.*, 260, 68, Jan. 1989.

Zaika, L. L. and Kissinger, J. C., Inhibitory and stimulatory effects of oregano on *Lactobacillus plantarum* and *Pediococcus cerevisiae*, *J. Food Sci.*, 46, 1205, 1981.

Zaika, L., Kissinger, J. C., and Wasserman, A. E., Inhibition of lactic acid bacteria by herbs, *J. Food Sci.*, 48, 1455, 1983.

Index